CONSTRUCTION MANAGEME

APPRAISAL, RISK
AND UNCERTAINTY

CONSTRUCTION MANAGEMENT SERIES

APPRAISAL, RISK AND UNCERTAINTY

Edited by
NIGEL J. SMITH

Thomas Telford

Published by Thomas Telford Publishing, Thomas Telford Ltd, 1 Heron Quay, London E14 4JD
URL: http://www.thomastelford.com

Distributors for Thomas Telford books are
USA: ASCE Press, 1801 Alexander Bell Drive, Reston, VA 20191-4400
Japan: Maruzen Co. Ltd, Book Department, 3–10 Nihonbashi 2-chome, Chuo-ku, Tokyo 103
Australia: DA Books and Journals, 648 Whitehorse Road, Mitcham 3132, Victoria

First published 2003

Also available from Thomas Telford Books
Financing infrastructure projects. T. Merna and C. Njiru. ISBN 07277 3040 1
Management of procurement. Edited by D. Bower. ISBN 07277 3221 8

Further titles in the series

Engineering management
Construction management
Operation and maintenance

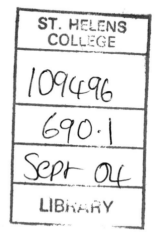
A catalogue record for this book is available from the British Library

ISBN: 0 7277 3185 8

Typeset by Helius, Brighton and Rochester
Printed and bound in Great Britain by MPG Books, Bodmin

Foreword

This second series of engineering management guides published by Thomas Telford aims to build on the success of the first series and provide knowledge relating to the current management issues facing the civil engineer in the 21st century. The first series, edited by Professor Stephen Wearne, consisted of a series of discrete, concise and practical guides to important management topics. Since their publication SARTOR'97 has radically changed the BEng and MEng curriculum and syllabi, and the practitioner has experienced new codes and regulations, BS 6079 project management, the Construction Design and Management Regulations, the Latham and Egan Reports, Engineering and Construction Contract, prime contracting, NHS 21, partnering, the Private Finance Initiative and the Turnbull report to mention just a few things.

Consequently, the new guides are intended to support graduate engineers and young chartered practitioners in the acquisition and effective management of fundamental knowledge relating to engineering management. The texts in the series are longer, integrated and designed to provide a sound basis for further reading or continuing professional development as appropriate. The first three guides to be published address the key aspects of Project Financing, the Management of Procurement and Appraisal, Risk and Uncertainty. These are relevant to all types of construction project. Each book reflects both the obvious interactions and dependencies between aspects of the provision and the need for specific knowledge at a strategic level. Further series guides will consider activities occurring during the project life cycle and will include Design Management, Construction Management and Operation and Maintenance. These are the more traditional engineering management topics presented at a tactical or project level including the dissemination of best practice. There will also be guides focusing upon discrete but significant topics such as Knowledge Management, Lean Construction, and Sustainability.

The new series editor is Nigel Smith, Professor of Construction Project Management, School of Civil Engineering, University of Leeds and is assisted by guide editors and by an editorial panel of the Engineering Management Group of the Institution of Civil Engineers.

Preface

Increasing pressures on the construction sector from owners and investors demanding improved project performance, while at the same time expecting the virtual elimination of waste and errors, has emphasised the need for a greater understanding of practicable project risk management. The increasing complexity of collaborative procurement methods and project financing, together with the existing problems of managing inherently risky projects to specification, on time and within budget, mean that engineers require a clear understanding not only of risk but also of the wider implications of appraisal and uncertainty management.

Appraisal, Risk and Uncertainty covers the basic terminology and the underpinning theory of risk management, the Institution of Civil Engineers and Faculty of Actuaries 'RAMP' methodology, the new and evolving aspects of corporate risk, the Turnbull Report, and the exciting possibilities of uncertainty management. The book is not concerned with trying to predict the future, rather, it is concerned with explaining how and when actions can be taken to support project decision-making under conditions of uncertainty.

The book is intended to be a guide to the engineer and to assist understanding. The chapters follow a chronological and developmental sequence, but can also be accessed at specific points as required. In many chapters, bibliographies are included for those seeking more specialist information.

Nigel J. Smith

Acknowledgements

The editor would like to thank Emeritus Professor Peter Thompson, UMIST, and Professor John Perry, University of Birmingham, for their permission to edit and re-present their publication on 'risk management in engineering construction', from 1986. This is included as Chapter 6, and is as relevant in 2003 as when it was first written.

The Institution of Civil Engineers and Faculty of Actuaries RAMP Working Party gave their permission to include an abridged version of their method as Chapter 7. The editor would like to thank all those involved, in particular Chris Lewin and Mark Symons.

The editor would also like to thank Cheryl Sonnier for all her assistance in converting draft text and artwork into polished products. However, any unknown errors remain the residual risk of the editor and authors .

List of contributors

Robert Ellis, BSc(Hons), MSc, PhD, FRICS, MCIArb ILTM, is a principal lecturer at Leeds Metropolitan University. Robert has experience both in local government and private practice, working as a chartered quantity surveyor on a variety of different projects. Recent research has focused on risk and value management, although he is also involved in several knowledge management commissions. Awarded a national teaching fellowship in 2003, Robert is currently developing innovative on-line teaching and learning resources.

Per W. Hetland, BSc, MBA, PhD, MNfP, NIFPDC, is a chief adviser for Statoil and an adjunct professor in project management at the Norwegian School of Management and at the Stavanger College. Per has extensive experience from major projects in the civil, mechanical, offshore and petroleum industries. Over the years he has played a major role in establishing organisations and programmes for enhancement of project management competence. He is a co-founder of NACPE (Norwegian Association of Cost and Planning Engineering), NfP (Norwegian Association of Project Management) and Epci (European Institute of Advanced Project and Contract Management). Per was the first president of NfP, and served several years on the boards of Internet/IPMA (International Project Management Association) and ICEC (International Cost Engineering Council).

Tony Merna, BA, MPhil, graduated with a BSc in Business studies from Salford University. He gained a Master of Philosophy degree from UMIST for his work on risk management in corporate organisations. Tony is currently undertaking a PhD programme at UMIST investigating the use of portfolio analysis for bundled projects, specifically to projects procured under a PFI strategy.

Nigel J. Smith, BSc, MSc, PhD, CEng, FICE, MAPM, is the Professor of Construction Project Management in the School of Civil Engineering, University of Leeds. After graduating from the University of Birmingham he spent 15 years in the construction industry, working mainly on transportation infrastructure projects. His academic research interests include risk management and procurement of projects using private finance. He is currently Dean of the Faculty of Engineering.

Gerard Wood, BSc(Hons), MSc(Cantab), MRICS, is a lecturer at the University of Ulster, specialising in construction economics and project management. He worked for several years as a chartered quantity surveyor in both private practice and management contracting, and his latest research has examined the practices of risk and value management in the construction industry. Gerard is currently undertaking an investigation of client and contractor experiences of partnering, sponsored by the RICS.

Contents

CHAPTER ONE

Risk and projects

N. J. Smith

This book is intended as a guide to the understanding of risk management and related techniques in the context of a construction project life cycle. It is not a specialised text on the mathematics of risk analysis but it is compatible with books of that nature. This book concentrates on the value of risk management as the process of making informed project decisions under conditions of uncertainty; it is not about forecasting, prediction or fate.

This first chapter introduces the common terms and definitions, and provides the context of the project life cycle and an awareness of project risk. A brief overview of project risk management is presented and the chapter concludes with an outline of the book contents.

Key definitions

The word 'risk' derives from the Latin *risicare*, 'to dare', clearly conveying the message that risk is a choice rather than a fate. At the end of the first millennium and beyond, many believed that the order of things was predestined and that nothing anyone did would change what was fated to happen. One of the fundamental concepts of risk management is that individuals and the decisions they take do matter and do make a difference. However, agreement on a precise definition of 'risk' is not as easy.

In the broader societal context, risk is defined and expressed in general terms:

- Douglas and Wildavsky (1981): 'risk is embedded in the same cultural values and norms that tell us what is right and wrong, what constitutes a democracy and what informs our political will'.
- Ansel and Wharton (1992): 'a measurement of the chance of an outcome, the size of an outcome or a combination of both'.
- Franklin (1998): 'there has always been a contingent edge to life and we use the term risk to talk about this contingency'.

However, there is a degree of agreement that in projects, risk is perceived as a potential adverse effect, as shown below:

- *Concise English Dictionary* (1976): 'the chance of hazard, bad consequences, loss etc.'.
- Lowrance (1976): 'a measure of the probability and severity of adverse effects'.
- Rowe (1977): 'the potential for unwanted or negative consequences of an event or activity'.

The more appropriate definitions of risk for construction projects are supplied by more specialised sources, including:

- HM Treasury (2000): 'the uncertainty of outcome, within a range of potential exposures, arising from a combination of the impact and probability of potential events'.
- BS 6079 (British Standards Institution, 1996): 'is the uncertainty inherent in plans and the possibility of something happening that can affect the prospects of achieving business or project goals'.
- Association for Project Management (2002): 'a combination or frequency of occurrence of a defined threat or opportunity and the magnitude of that occurrence'.
- Smith (2002): 'risk is adverse but an unknown by its nature can have both positive and negative effects'.

This cluster of definitions provides a better understanding of the nature of project risk than any single definition. To paraphrase and combine them, it is clear that there are sources of risk which can be assessed by considering their probability of occurrence and their adverse impact on the project objectives, and there are genuine unknowns whose outcome could be beneficial or detrimental to the project objectives.

The term 'uncertainty' is used in a variety of ways in risk management literature. In recent years it has become associated with 'uncertainty management', which is the process of integrating risk management and value management approaches and is discussed in detail in Chapter 8. Here the more traditional definition can be traced back to the 1920s (Knight, 1921):

> The practical difference between the two categories, risk and uncertainty, is that in the former the distribution of the outcome in a group of instances is known ... while in the case of uncertainty this is not true, the reason being in general that it is impossible to form a group of instances, because the situation dealt with is in a high degree unique.

More recently, Chapman and Ward (2002) have defined uncertainty as 'a lack of certainty; involving variability and ambiguity' and uncertainty management as 'managing perceived threats and opportunities and their risk implications but also managing the various sources of uncertainty which give rise to and shape risk, threat and opportunity'.

This text follows the Chapman and Ward definitions for general uncertainty and a specific task of uncertainty management. The use of 'uncertainty' in the Knight quote given above is similar to the use of 'unknown' in this text.

The third term from the title of the book is possibly the most understood and yet most difficult. This is because the term 'appraisal' has been used for two different things, namely an evaluation of something and also the first phase of a project life cycle. Appraisal, the technique, takes place in the appraisal phase of the project. These differences are explored in detail in Chapters 2 and 3.

An appraisal is an evaluation of something, according to Gower (2000): 'the assessment of the benefits and the feasibility of the lessons learned within the organisation'.

The need for risk management

Risk has been a topic of continued management interest since projects were first undertaken. Some general guidance is available on appraisal or value management, but when these subjects are raised they are treated more as discrete and individual techniques or processes rather than as complementary activities in the early dynamic stages of project management. The whole ethos of these new guides is to promote integrated or 'joined-up' thinking in a management approach to construction projects. In line with this philosophy this book will offer one of the first integrated approaches to all aspects of risk.

In the real world all projects are undertaken under conditions of some level of uncertainty. Very few things are known with complete certainty and it is increasingly unlikely that it will be possible to control all events on major, complex projects. Most projects require an implementation or construction phase and, as stated in the Royal Charter of the Institution of Civil Engineers of 1828, this may involve 'Harnessing the forces of nature for the benefit of mankind', which is inherently risky.

Understanding the nature and importance of risk and risk management is not helped by the inconsistent use of terms. This book will use the terminology consistent with project risk, as defined in BS 6079. The very nature of risk management can cause difficulties in understanding or perceiving exactly what the purpose of the exercise really is! This is a good opportunity to dispel one or two myths about risk management and establish some core facts. Risk management is not about predicting the future, nor is it a single, one-off calculation fixed for the duration of the project. Also, the execution of a risk analysis does not of itself change anything in the real-world project. Rather, risk management is about communication to make better decisions on a real project under conditions of uncertainty. It is a continuous and dynamic process which is necessary from project inception until project conclusion, although what tends to be thought of as conventional risk management is undertaken in the early stages of a project.

If risk management was simply a question of learning by mistakes, once all the failures had occurred in a system it would be possible to eliminate all future failures. It is obvious that this is not realistic.

Project viability

There are usually many more ideas for projects than there are viable projects. Hence one of the fundamental roles for risk management is to identify and distinguish between viable and non-viable projects. There are benefits to all parties in identifying non-viable projects as early as possible; however, this conflicts with the fact that at the early stages of a project many uncertainties exist and decision-making is difficult.

Determining the viability of a project is the first decision made in the appraisal phase of the project and is explained in Chapter 3.

BS 6079 and HM Treasury

There are a range of guidebooks, codes of practice and project protocols that can be utilised to assist the risk management process. One process, the RAMP approach developed in conjunction with the Institution of Civil Engineers, is described in Chapter 7. However, there are two important documents that have influenced the presentation and content of this book, namely the British Standard BS 6079 and the HM Treasury guides.

In 2000 in the UK, the third part of the British Standard on project management, BS 6079, was issued on CD and in hard copy; it includes definitions, tools and procedures for risk management. Although not directly concerned with risk, it provides a valuable project context in which the processes of risk management can operate.

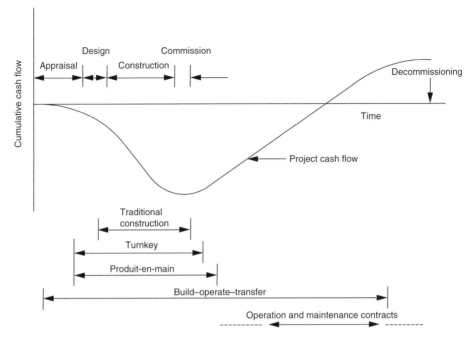

Figure 1.1. Project cycle

For many years HM Treasury has been producing guidance notes on the standards expected in the award and administration of government contracts. Recently these were issued by the Central Unit on Procurement but in 1997 the CUP guides were superseded by the current documents issued by the Procurement Group. Guide Number 2, *Value for Money in Construction Procurement*, contains a useful appendix that includes guidance on risk management for public sector projects. These give a clear understanding of what and how contractors interested in government work have to perform. The basic approach is compatible with the RAMP Guide approach described in detail in Chapter 7. Nevertheless, UK practitioners should be aware of both sets of documentation.

Risk in the context of the project

The project cycle is a useful vehicle for considering the dynamic nature of risk assessment. Figure 1.1 shows the cycle and the three main phases: appraisal, implementation and operation.

BS 6079 defines a project as 'a unique set of co-ordinated activities, with a definite start and finish point, undertaken by an individual or organisation to meet specific objectives with defined schedule, cost and performance parameters'. Turner (1999), explains that projects are undertaken to fill the planning gap between where an organisation would like to be and where it predicts it will be if nothing is done. However, it is also important to remember that in many cases the 'do-nothing' option is better than the project proposal. Risk management has a role throughout the project life cycle but a disproportionate part occurs during the appraisal phase. The reasons for this are explained in Chapters 2 and 3.

Book outline

The second and third chapters concentrate on the appraisal phase and the project appraisal process itself. This is the most important phase of the project life cycle, where changes can be cost-effective and have a major effect on the project, although by definition this phase has the maximum uncertainty associated with the project.

Chapter 4 provides a useful overview of the range of existing approaches to the management of risk in a civil engineering context. The chapter also considers how some of these approaches are managed in practice and investigates workshops and brain-storming sessions.

The classification or categorisation of risk can be helpful for iden-tification purposes and also for assisting the decision maker in deter-mining priorities. There is no set or standard methodology, but two approaches are outlined in Chapter 5 and some checklists included.

The foundation for the currently accepted approach to risk management, first conceived over 30 years ago and revised over 10 years ago, is still valid. Chapter 6 reviews the methodology of risk identification, risk assessment and residual risk management. Leading on from the basic theory, Chapter 7 is devoted to a précis of the Risk Analysis and Management of Projects (RAMP) approach, developed jointly by the Institution of Civil Engineers and the Faculty of Actuaries.

Uncertainty management is the subject of Chapter 8. An excit-ing and interesting view of the future of uncertainty management is presented, and cross-comparisons are made with the RAMP approach. This aspect of risk management is the subject of ongoing development and has interest for most current practitioners.

Another recent development has been the changes in the approaches taken to the management of and responsibility for corporate risk. Chapter 9 commences with a review of the Turnbull Report, which proposed the new methods, and then links between corporate risk and project risk.

Finally, Chapter 10 briefly looks at recent trends in risk management and speculates on possible future developments.

Bibliography

Ansell, J. and Wharton, F. *Risk: Analysis, Assessment and Management.* Wiley, Chichester, 1992.

Association for Project Management. Website: www.apm.org.uk.

British Standards Institution. *Guide to Project Management.* BSI, Milton Keynes, BS 6079, 1996.

Chapman, C. and Ward, S. *Managing Project Risk and Uncertainty.* Wiley, Chichester, 2002.

Concise English Dictionary. Oxford University Press, Oxford, 1976.

Douglas, M. and Wildavsky, A. *Risk and Culture.* University of California Press, Berkeley, 1981.

Franklin, J. (ed.). *The Politics of Risk Society.* Polity Press, Cambridge, 1998.

HM Treasury. *Management of Risk – A Strategic Overview.* [Orange Book.] HMSO, London, 2000.

Knight, F. H. *Risk, Uncertainty and Profit.* Houghton Mifflin, Boston, 1921.

Lowrance, W. *Of Acceptable Risk: Science and Determination of Safety.* Kaufmann, Los Altos, 1976.

Rowe, W. D. *An Anatomy of Risk.* Wiley, London, 1977.

Smith, N. J. *Engineering Project Management,* 2nd edition. Blackwell, Oxford, 2002.

Turner, J. R. *The Handbook of Project-based Management,* 2nd edition. McGraw-Hill, Maidenhead, 1999.

Turner, J. R. and Simister, S. K. *Gower Handbook of Project Management,* 3rd edition. Gower, Aldershot, 2000.

CHAPTER TWO
The project appraisal phase

N. J. Smith

Appraisal is the term used to identify the early phase of the project life cycle. However, as shown in Chapter 3, it can also be used to describe a process of quantitative and qualitative evaluation. This chapter reviews the appraisal phase of the project in detail, including the significance of this phase for the risk management process.

Definition

Project appraisal is the early stage of the project cycle, and commences with the project inception and finishes at project sanction. This phase can be subdivided into a 'pre-feasibility' or 'viability' stage and a 'feasibility' stage. Appraisal is an important stage in the evolution of any project because at the early stages of a project it is possible to make changes that are relatively cost-effective. It is important to consider alternatives and identify and assess risks, at a time when data are uncertain or unavailable.

Appraisal in the project life cycle

The basic project life cycle, as shown in Figure 2.1, consists of three main phases: appraisal, implementation and operation. These phases cover the cycle between inception and sanction, between sanction and commissioning, and between commissioning and decommissioning, respectively.

Within the appraisal phase of the project there are two different types of decision process to be managed, both of which are determined under conditions of uncertainty. The initial decision is often referred to as the 'viability' decision, which requires the justification of whether the project is realistic and whether further investment in the project should be made. If the decision is that the project is not viable then there is no second decision to make, but if the project is found to be viable then the second decision is the 'feasibility' decision. Given that the outline concept for a project is viable, the

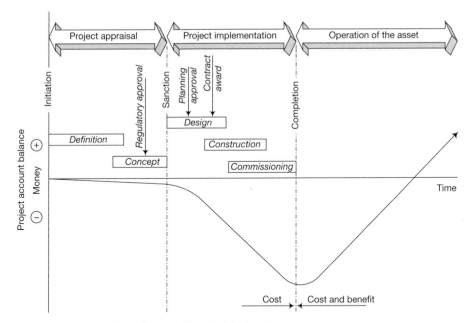

Figure 2.1. Project life cycle: cash flow (Smith, 1999)

feasibility decision is to identify the particular form or option for the project out of all possible alternatives that is likely to be the most successful.

Viability

The decision regarding viability is in one sense a simple 'yes' or 'no' choice; however, it is also a very difficult decision because whilst rejecting a poor project at an early stage could save an organisation a large amount of money, rejecting a potentially very profitable project at the same stage could be disastrous for the organisation. The work undertaken has to be sufficiently detailed and requires sufficient investment to make a good decision but, equally, must try to avoid wasting expertise and investment on overexamining a non-viable project. The decision process is further complicated by the fact that this must take place as soon as possible after the inception of the project, which usually coincides with the time of maximum uncertainty about the project.

Feasibility

At the second decision stage, the feasibility stage, the decision is to identify the most appropriate option for the project from amongst

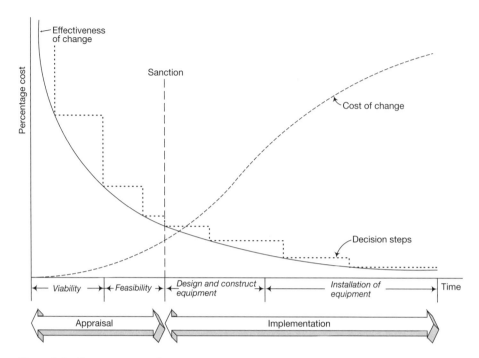

Figure 2.2. Percentage cost plotted against time, showing how the important decisions for any project are made at the start of that project (Smith, 2002)

all the possible alternatives. The process is often referred to as a feasibility study. As the project progresses through the appraisal phase the range of uncertainty is reduced as project options are either confirmed or eliminated, although at this stage the process still involves decision-making under conditions of uncertainty.

There is no standard method for making this decision, and a wide range of tools and techniques are available to assist the project manager by providing better information on which to base a decision. No matter how good the computer software or how rigorous any analysis, the final decision has to be made by the project manager within the overall strategy and goals of the project.

Related processes and applications

Risk management in general and the appraisal process in particular are not helped by confusion over terms and definitions. Although there are no substantive differences regarding the composition of the project life cycle, the terminology used by different industrial sectors varies widely and in some cases uses the same term to identify a different phase of the cycle. Figure 2.2, based on Fine (1982), illustrates the relationship between the cost of making a change, the impact of the change in terms of project savings and the timing of the change. Essentially, all major changes need to be made during the feasibility or appraisal stage of the project, i.e. prior to project sanction. However, this is precisely the stage of the project at which uncertainty is at a maximum.

Although most textbooks concentrate on individual projects, it is likely that for most organisations a single project will only constitute part of corporate business. It is more likely that there will be several alternative projects competing for the available resources. A choice or prioritisation of projects has to be made.

The cash flow changes with the project phase. During appraisal, a relatively small amount of money is invested to fund the work of the project management team. After sanction, large numbers of organisations, typically contractors, suppliers, vendors, subcontractors, specialists and others, are engaged and the major investment in the project occurs. Typically, although not in every case, the maximum investment in the project, also known as the capital lock-up, occurs at commissioning, i.e. the completion of the implementation stage; a typical investment curve is shown in Figure 2.1. Once commissioned and operational, the project starts to generate revenue, which pays off debt and interest until payback, the date at which the project breaks even. After payback the project starts to generate profit but must cover the costs of operation and maintenance.

Risk management in the appraisal stage

Project appraisal is a process of investigation, review and evaluation undertaken as the project or alternative concepts of the project are defined. The techniques and tools are discussed in Chapter 3, but it is the timing of the decision-making within the appraisal phase that is considered next.

The basis for this discussion is shown in Figure 2.2. The horizontal axis is time-based and shows the appraisal stage and the implementation stage. The vertical axis reflects change in two ways, firstly in terms of the effectiveness of a change and secondly in terms of the cost of making the change. Two curves are shown in the figure, a solid curve representing the effectiveness of change over time and a dotted curve representing the cost of change over time.

The solid line falls rapidly during the viability substage and then falls less steeply across the feasibility substage, whilst during the same period the cost curve rises slowly from zero. This is not surprising, as during the viability stage very little detail is known about the project, and hence changing ideas or concepts is cheap and can have significant effects upon the project. As both curves progress beyond the sanction stage, the effectiveness of a change decreases until it is of very little value, and the costs of change increase hugely. This reflects the reality of the project, where decisions have been made, contracts have been awarded and hence the scope for change is reduced and the cost of making the change probably involves contractual issues. In practice, the money invested in a project during appraisal is somewhere between 5% and 10% of the capital lock-up for the project. Decisions made at the sanction stage and in awarding contacts will freeze about 80% of the remaining cost, hence the shape of the curves in Figure 2.2, showing that the opportunity to reduce cost during the implementation stage is small.

Summary

This chapter has reviewed the appraisal phase of a project and has investigated the substages of viability and feasibility. Knowledge of this stage provides the reasoning for conducting risk management at the earliest stages of a project. Although risk management covers the entire project life cycle, the major input and effectiveness are in the work conducted during the appraisal stage, where it is cheaper to make changes and the changes are likely to be more effective for the project as a whole.

Bibliography

Fine, B. *Journal of the Association of Project Managers,* **12** (1982), 2–5.

Smith, N. J. *Managing Risks in Construction Projects.* Blackwell, Oxford, 1999.

Smith, N. J. *Engineering Project Management,* 2nd edition. Blackwell, Oxford, 2002.

CHAPTER THREE

Decision-making during the appraisal phase

N. J. Smith

This chapter outlines some of the many processes used to assist decision-making during the project appraisal stage and explains its role with respect to project risk management. The nature of decision-making at both the viability and the feasibility stages of a project is reviewed and the need for a reliable method for assessing alternatives is established. The chapter concludes with an outline of how the appraisal phase of a project is typically managed.

Appraisal terminology in sector-specific project life cycles

The industrial terminology for the substages in the appraisal stage of a project is not uniform. Figure 3.1 (McGettrick, 1996) includes a number of examples of industrial-sector-specific project life cycles and demonstrates the need to define terms at an early stage in any project to avoid ambiguity. The table shows clearly that most sectors recognise the division of the appraisal stage, although the terms 'viability' and 'feasibility' are not always used.

There is also a range of terms for the tools and techniques used for project evaluation and assessment. Once again care is needed to ensure that communications are clear and that misunderstandings are avoided.

The appraisal decision

As outlined in Chapter 2, the appraisal phase is important for making a series of decisions, firstly the viability decision, secondly the feasibility decision and thirdly, on the basis of the output of these decisions, the sanction decision.

In order to make a decision the project manager must know 'what the project is aiming to achieve', that is, the project objectives must be clear and coherent. If there is any doubt as to the importance or priority of project aims it will be difficult to make effective decisions.

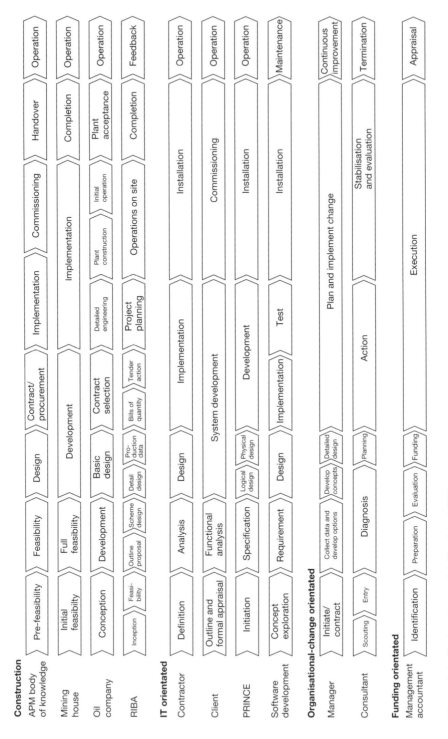

Figure 3.1. Project phases (McGettrick, 1996)

Decision-making is critical to the success of the project; at viability a yes/no decision is made, whilst at feasibility a key decision on the selection of the project option is required. During the appraisal phase the degree of certainty about the project increases. By the sanction decision point about 80% of the cost of the project will be frozen, which offers little scope for additional time or cost savings. After sanction, it is usual for contracts to be signed and for the implementation stage to proceed at the appropriate quality standards, with a realistic time schedule and budget.

For many large construction projects, particularly those overseas in remote areas, the range of risk and uncertainty can be underestimated. This is partially due to the degree of uncertainty associated with the project and partially due to a tendency for people working on a project wanting the project to succeed. Hence all parties involved in construction projects would benefit greatly from reductions in uncertainty.

Owing to a lack of project definition, the appraisal stage should focus on seeking solutions which avoid risk, trying to evaluate the overall riskiness of the project and to assess the implications of the evaluation in terms of decisions made at sanction, including procurement strategies.

Appraisal uses the project life cycle. For many projects the life cycle is over long periods of time, and using discounted cash flow techniques tends to distort values. For example, in offshore oil fields there is often a large cost for decommissioning of the platform, which occurs at the end of the life cycle. For other projects it may be necessary to consider the development in terms of 'what if' scenarios. The scenarios may be small in number and test key project options such as delays or production/revenue rates. Sometimes these can be expressed by calculating a spectrum of outputs on the project cash flow cycle. An example is given in Figure 3.2.

It is important to remember, from the shape of the investment curve, that interest payments compounded over the project duration until payback will form a significant element of project cost and the investor will not derive any benefit until the project is completed and is in use.

The viability decision

The earliest of all possible project substages, viability, commences at inception. There are always more project concepts or ideas than there are viable projects. Therefore some assessment or evaluation of the project needs to be undertaken at this substage, despite the maximum level of project unknowns.

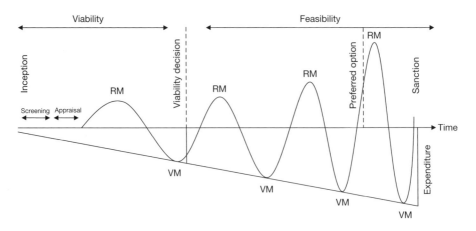

Figure 3.2. Decisions in the appraisal phase

There is no standard methodology for this process of decision-making, and different tools and techniques are utilised by different organisations. However, there are some practices commonly used by a range of organisations and practitioners. A typical starting point is 'screening' or 'filtering'. This is a method of working through a checklist of factors, any of which is likely to put the viability of the project in question, for example:

- Does the project complement the organisation's strategic objectives?
- Is the budget affordable?
- Is the location politically suitable?
- Has the organisation any experience of this technology?

Depending upon the degree of development of the system, these questions may require simple yes or no answers, or receive a relative score on a Likert scale, or form part of a more sophisticated decision matrix or decision table. For some organisations it is possible that a single 'no' answer is sufficient to eliminate the project idea.

If the project passes the screening process then it may be possible to conduct a base case appraisal, which is a first-order appraisal assuming that no risk occurs. In order to prepare this representation of the project, the size, timescale and major elements of the project must be identified. Again this will be easier for some types of project than for others. At this stage the same information can assist in the derivation of the first budget. All these estimates and predictions, which frequently extend over long periods of time, will generate different degrees of uncertainty.

At these early stages it is often difficult to obtain realistic cost data. However, wherever possible a basic cost–benefit analysis should be attempted to reflect opportunities and social benefits. No factor should be ignored because it is difficult to quantify in monetary terms. Methods are available to express, for instance, the value of recreational facilities, or the value of disability through injury, although the actual figures may not be accepted by all parties.

For the viability decision, the usual test is to compare the cost–benefit analysis with the available budget and future income generation options. As the level of certainty increases throughout the appraisal stage, this budget may form the basis for a more informative investment appraisal and possibly, by sanction, a full business plan for the project. For some projects that will be the information on which the viability decision is based; for others there may be the opportunity to use the base case to generate a first risk management and/or value management assessment. The process of risk management is covered in detail in the later chapters of this book and there are many texts on value management, including a guidebook in the current series.

The feasibility decision

Once the viability of a project has been demonstrated, the next decision is to identify the most appropriate project option. This option will then form the basis for the major decision to be made at sanction: whether to invest in the project or not. Many projects are not sanctioned, and although this is more expensive than making a decision at the viability stage, only about 10% of the project value has been invested and the remaining 90% can be utilised on better project options. This is obviously a key decision in the life cycle of the project.

There will always be conflict between the desire to gain more information and thereby reduce uncertainty, and the need to avoid wasted expenditure on a project appraisal that will have to be written off if the project is not sanctioned. Some expenditure must be made if informed and rational choices about the project options are to be made. Appraisal is likely to be a cyclic process repeated as new ideas are developed, additional information received and uncertainty reduced, until it is possible to make a reasoned choice between project options.

Ideally, all alternative concepts and ways of achieving the project objectives should be considered. The resulting proposal prepared for sanction on the basis of the best project option must define the major parameters of the project – the location, the technology to be used, the size of the facility, and the sources of finance and raw materials, together with forecasts of the market and predictions of the cost/benefit of the investment.

At sanction, the project option will again be assessed against the project objectives, which may have been refined during the appraisal process. Many projects are undertaken for commercial business ventures and information about future markets can be particularly important at this stage. Often the first producer to market has a bigger impact and higher sales, and hence market intelligence might cause a substantial change to the project objectives.

Since the viability decision, the information to feed a risk analysis will have been improving and the analyses will have been repeated at regular intervals during the appraisal phase. The findings from the analysis will help to establish confidence in the project estimate and procurement strategy and to allocate appropriate contingencies. This should provide a basis of information that can be used to determine the most effective way to implement the project and to achieve the project objectives, taking account of all constraints and risks.

Non-monetary appraisals

Finally, dependent upon the type of project being executed, there are a number of other types of appraisal that will often be carried out during the project appraisal stage. These appraisals are relatively well known and are becoming increasingly important for construction projects in the 21st century. Four of the most common are briefly described below:

- *Environmental appraisal.* This is a key element in most large project appraisals, considering impacts and mitigation measures on the local environment during implementation and operation. Control of pollution from process plants and the management of waste products has become highly regulated in the UK in recent years.
- *Health and safety appraisal.* Apart from the general responsibilities under statute, i.e. the Health and Safety at Work Act, and under contract law, the CDM regulations place restrictions on all designers to ensure that there is a safe method of erection of the design. This is not always the method finally adopted on the site.
- *Ethical appraisal.* As international and multicultural working become more common, the need for ethical awareness is increased. Investors are often selecting organisations with ethical development plans in preference to those organisations that may be associated with the arms industry or child labour in developing countries.

- *Sustainability audit.* This is a key process for all development which takes place on a planet with finite resources and increasing quantities of waste materials. Most audits adopt the 'Bruntland' definition of 'providing the services for today without damaging the prospects for tomorrow'.

Typical risk management in the appraisal phase

Despite the heading of this section, it is difficult if not impossible to define a typical project or typical risk, and the precise actions undertaken on any real project would be governed by the level of understanding and the requirements of that project. Nevertheless, there is a rational approach to the sequencing of activities in the appraisal stage and this is presented below.

The key events are illustrated in Figure 3.3. It can be seen that risk is a dynamic and continuous process which is reiterated throughout the appraisal stage and beyond. Risk management should never be regarded as a single evaluation; it will change with the degree of certainty and the decisions made as a consequence of the project progressing.

The outputs from the risk analysis will play a major role in the feasibility process. The cumulative frequency curves will give an indication of both the inherent riskiness and the profitability of each option. However, it will be necessary for the project manager or a similar professional to make the actual decisions at sanction on the basis of this detailed information.

The risk analysis will also be used to determine the duration of the appraisal and implementation phases and to help decide on the procurement strategy. The importance of time should be recognised throughout the appraisal. Many costs are time-related and would be extended by any delay. The programme must therefore be realistic and its significance taken fully into account when determining the project objectives. Similarly, major unquantifiable risk would usually indicate the selection of a cost-based procurement strategy and a low-risk project would favour fixed-price lump sum contracts.

Summary

This chapter has reviewed the decision-making processes in the appraisal stage of a project. It found that there are three effective decisions to be made in sequence. Initially, there is the viability decision, which is designed to check if the project idea or concept is viable. Then there is the feasibility decision, at which the most

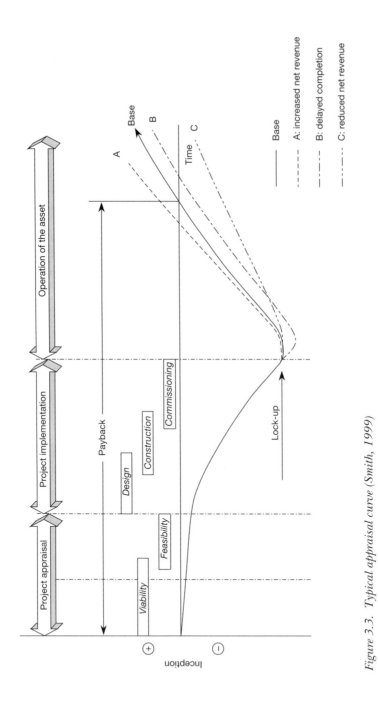

Figure 3.3. Typical appraisal curve (Smith, 1999)

appropriate project option should be identified. Finally, there is the project sanction decision, at which investment is committed to the implementation of the project. Through the appraisal stage, risk analysis plays an increasingly important part in helping to inform the decision maker with the best data possible on which to base the decision.

Further, this chapter, in conjunction with Chapter 2, showed that the appraisal phase is the phase where the major risk management has to be undertaken, because once the sanction stage is reached it is very difficult to make effective changes and increasingly expensive to try to do so. Hence the need to concentrate on the early stages of the project cycle.

Bibliography

ICE Guide on Value Management. Thomas Telford, London, forthcoming.

McGettrick, S. The project life cycle – interpretation. *Project*, June (1996), 13–15.

Male, S. P., Kelly, J. and Graham, D. *Value Management of Construction Projects*. Blackwell Science, Oxford, 2003.

Smith, N. J. *Managing Risk in Construction Projects*. Blackwell Science, Oxford, 1999.

CHAPTER FOUR

Existing risk management approaches in civil engineering

G. Wood and R. Ellis

Introduction

Since the mid-1980s many authors have suggested that the management of construction projects, large or small, benefits from a greater understanding brought about by the application of risk management techniques. The well-known work of Perry and Hayes (1985) established risk management as a concept of relevance to construction projects and elaborated upon a three-stage process that comprised identification, analysis and response. Perry and Hayes concluded that risk and uncertainty were not the sole preserve of large capital projects but that factors such as complexity, speed of construction and location also contributed to the inherent risk within a project. Risk management is now widely accepted as a vital tool in the management of projects, and in recent years an array of documents have been published which aim to provide guidance for practitioners undertaking the risk management process. Typical examples include CIRIA (Godfrey, 1996), HM Treasury (1997), the Association for Project Management (1997), the Institution of Civil Engineers *et al.* (1998), the British Standards Institution (2000) and the Royal Institution of Chartered Surveyors (2000).

Simultaneously, there has been an increase in research aimed at investigating risk management practice in the construction industry, such as the work of Simister (1995), Potts and Weston (1996), Akintoye and MacLeod (1997), Jackson *et al.* (1997) and Amos and Dent (1997). However, Edwards and Bowen (1998) question the adequacy and appropriateness of what has been largely opinion-survey research (based upon questionnaires) and suggest that methodological weaknesses undermine the validity and usefulness of the findings. They advocate the greater use of case study techniques using the real experiences of project participants to examine important soft systems issues. Hence this chapter draws heavily on the qualitative work of Ellis and Wood (2001) and Wood and Ellis (2001), which explores current attitudes and experiences of risk management facilitators.

Client demand

The Royal Institution of Chartered Surveyors (2000) suggests that major clients have formal procedures in place for managing risk, but that other clients are likely to seek guidance from their professional advisers on how best to deal with risk. Indeed, a broad range of clients, for example developers, pharmaceutical and oil companies, banks, lawyers, insurance brokers, government organisations, transport, and utilities, now commission risk management studies. The latter, in particular water, gas and power, incorporating transmission and generation, adopt rigorous procedures as a matter of routine, where risk management is accepted as being part of the company's culture. One of the central issues for consultants, though, is not whether risk management is perceived by their clients to be of value, but whether they are prepared to pay for risk management as an additional service.

For clients within public sector domains such as transport and infrastructure, quantitative risk analysis is undertaken on each project. Conversely, industries associated with small-scale projects are less likely to commission risk management. However, the relative strength of interest expressed by the private sector is variable. The reason for this apparent dichotomy may be due to the lack of need for public accountability, or simply because private sector clients trust the advice given by their consultants and believe they implicitly take risk into account without a formal procedure to control risk. A more cynical view might be that consultants' professional indemnity policies enable clients to recover losses should schemes run into trouble.

The nature and complexity of the project are therefore key factors that influence the decision to undertake a risk management study. Railtrack and BNFL are cited as companies operating in complex, high-risk environments where risk management is widely regarded as being essential. However, it may be erroneous to generalise service requirements by reference to a particular industry. Although risk management could be part of company culture, quite how the service is implemented tends to be manager specific. Seemingly, this is more prevalent in the water utilities than in power and gas transmission, where risk management is applied more broadly throughout the whole life cycle.

In general, risk management is perceived to be a growth area and the indications are that it will continue to expand. For example, Private Finance Initiative (PFI) schemes actually require risk management studies to be undertaken, as do lottery-funded capital projects. In essence, PFI contracts can be seen as a means of unloading cost risk and product performance risk onto the private sector. Consequently, contractors often seek reassurance or comfort related to their risk exposure on PFI schemes by undertaking a risk management study. In addition, some funding organisations, particularly banks, now insist on risk management procedures being undertaken in order to mitigate their risks. It therefore appears that there may be a shift in the nature of risk management services as more clients demand that a full business perspective is taken. The Turnbull Report (Institute of Chartered Accountants, 1999), for example, has been influential in raising the profile of risk management generally in that directors of publicly quoted companies are now required under Stock Exchange regulations to consider risk in the operation of their companies.

Implementation within the project life cycle

Perry and Hayes (1985) recognised that the greatest degree of uncertainty is encountered during the early stages in the project life cycle and that the cost implications of decisions made at this time would have a dramatic effect on the overall viability of the scheme. It is not surprising therefore that a risk management consultant should want to be involved as soon as possible. As a result, most studies are undertaken during the conception phase, where clients are seeking to evaluate and compare different options. Yet for many, this is not soon enough. Risk management should ideally commence at the pre-design phase when high-level strategic decisions are made which affect overall business development and procurement strategies. Rather worryingly, the enthusiasm for risk management from a client perspective demonstrated during the initial phases of a project is seldom sustained throughout the whole life cycle.

Commissions received, particularly in the private sector, often do not continue through to construction. However, this is not universally the case, for although risk is generally perceived to have been handed over to the contractor by the post-contract stage, much depends upon the chosen procurement option. Risk registers may change perhaps every three to six months and the need to revisit mitigation and contingency plans does not necessarily diminish simply because the contractor is on site.

Service delivery

Commonly, the nature of a risk management commission is analogous to the shape of a diamond (Figure 4.1). The initial inquiry,

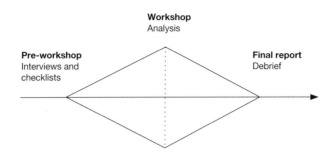

Figure 4.1. Consultant involvement on a risk management commission

which generates contextual information, is followed by a series of intensive interviews and workshop sessions. Some analytical work is undertaken at the workshop stage in order to develop a risk register and identify those members of the project team who are to be held responsible for mitigating the effects of the risk. However, the bulk of the analysis takes place after the workshop sessions and prior to the issue of a draft report. In addition, some consultants conduct feedback meetings with key project stakeholders and attempt to close out issues. Occasionally, as part of an ongoing involvement, risk managers monitor whether the mitigation plan is being implemented.

What, then, does a risk manager seek to manage? A distinction must be made between managing each risk and managing the process. This is perceived to be a subtle yet important distinction because professional indemnity insurance does not necessarily cover the consequences of a risk occurring, and the risk must be managed by the party most able to manage the risk. A broad spectrum of risks is usually considered, for example buildability, construction, health and safety, logistics, business continuity and political risks. Typically these risks are expressed in cost terms rather than in terms of the effect on programme and time schedules.

The consensus view is that blanket contingencies (say, plus 10%) are to be avoided and that risks have to be identified and priced in order to determine an accurate and realistic budget. Occasionally this might be extended to a consideration of life cycle costing but rarely would the risk manager provide detailed advice on corporate finance, etc., preferring to leave such matters to the client's accountants.

The risk management process

The most effective forum for the early stages of risk management is the workshop. Almost all practitioners attempt to convene a risk management workshop attended, where possible, by all major stakeholders: clients, project managers, designers, cost consultants, contractors (where appointed), end-users and even in some instances external organisations such as a local residents' association. However, there are often important tasks to be undertaken pre-workshop.

Where time permits and the scale of the project allows, many consultants prefer to interview stakeholders prior to the workshop. The intention here is to get a general feel for the principal concerns of those involved in a project. This is seen by some as a crucial part of the process because of the improved quality of information obtained. When interviews are not possible, consultants often send out briefing papers to introduce the risk management process and explain its aims and objectives. In addition, some send out questionnaires or round robins as a means of obtaining at least some impression of stakeholders' views prior to the workshop. These questionnaires are sometimes referred to as risk identification forms and provide the basis for a first-draft initial risk listing which can be tabled at the workshop.

The length of time dedicated to workshops tends to range from a half day to two days depending upon the nature of the project and the willingness of the client to pay given that a two-day event is estimated to cost in the order of £10 000. The most common duration is a half or full day.

The workshops themselves tend to follow a similar format, involving either group or individual brainstorming and Delphi techniques to identify risks initially. Facilitators use prompt lists to direct and stimulate the group's thinking and subsequently they employ checklists to ensure all issues have been aired. The workshop is used to rank the identified risks, and most go on to use some kind of scoring system to take account of both the probability of the risk occurring and the consequent impact on the project (see the later section on risk analysis techniques). Brainstorming sessions within the workshop encourage lateral thinking, and the ranking and scoring process captures the collective intelligence of the project team and achieves a consensus on what are the major project risks. In addition, the workshop environment can provide an opportunity for valuable interfacing between all stakeholders, opens up channels of communication and can even become a team-building event. However, the views of participants can remain hidden in the workshop situation. For instance, if certain members of the design team are perceived by others to be a major contributor to risk, it is unlikely that this would surface at a round-the-table meeting. Hence the preference of some for individual pre-workshop interviews.

Following the workshop, further evaluation of the issues is carried out and, in the majority of cases, a project risk register is published which identifies individuals to take ownership of each particular risk. An accompanying risk report for the client including mitigation plans is also produced. Frequently, there are follow-up meetings with the project team, typically of one to two hours' duration, and in some cases there are further review meetings at certain project milestones, for example, at the letting of the contract, when a number of risks may be passed over to contractors.

Classifying risk

There are a variety of approaches to categorising the identified risks. Perhaps the most common is to use the origin or consequence of the risk, for example programme risks, cost risks, site risks, etc., or to use specific construction features such as structure, service installations or even a full BCIS-type elemental breakdown. Some practitioners classify risks according to the stages within the development process along the lines of the RIBA plan of work – design development, specification, procurement, tendering, construction, operation and maintenance. On larger projects, risks are often grouped either within different phases or in line with the project work breakdown structure. Other practitioners take a broader view by using categories such as political, environmental and commercial risks or by focusing on capital costs, maintenance costs and life cycle costs. Such methods are really seen as ways of helping the team to consider risks across the whole project and are not used as a rigid framework. The differentiation of risk and uncertainty and the classifications of risk offered by some texts such as dynamic, static, pure, speculative, controllable and uncontrollable are hardly ever adopted in practice. This bears out the suggestion by Perry and Hayes (1985) that such distinctions are usually unnecessary and may be unhelpful.

Risk analysis techniques

Checklists from previous projects are used to help identify potential risks on new projects and occasionally practitioners create their own database of risks. This is seen as particularly useful where the projects are of a similar type or with a repeat client. However, the majority of information comes from the project team during the workshop(s). The risks and values drawn out of the workshop participants are then interrogated, to some extent, by the rest of the team. This is considered to be a means of ensuring the values are justified and goes some way to testing data for bias. With regard to the actual figures and values used in a risk assessment, there is often only limited use of historical or out-turn data from previous projects.

A probability–impact matrix is regularly used, at least initially, to help assess, rank and score the identified risks. Scoring systems range from simple 'low, medium, high' ratings to more precise numeric scales such as those suggested in the RAMP guidebook (Institution of Civil Engineers *et al.*, 1998), where both probability and impact are scored on a calibrated scale and one is multiplied by the other to give a combined value. The use of matrices is thought to help focus the minds of the team on what they have to manage and where the emphasis ought to be. Few consultants use decision trees, fault trees or event trees. Similarly, there is little use of spider diagrams with or without probability contours, influence diagrams, linear regression, torpedo models, neural networks or other more complex techniques. Sensitivity analyses generally comprise a straight-forward quantum test of the overall impact of changes in some of the variables.

Almost all practitioners, however, use specialist computer soft-ware to calculate minimum, maximum and most likely values of each risk or combination of risks using Monte Carlo simulation. Amongst the most common packages are @Risk™ and Predict™. Whilst there might be a good general understanding of the princi-ples adopted in simulation modelling, there is perhaps less confi-dence regarding the detailed workings of the model, for example the difficulties associated with the potential interdependence of variables or the selection of an appropriate distribution profile for each risk. Commonly, only a three-point triangular distribution is used, with perhaps some use of normal and uniform distributions. The reason for this is that it is often simply a matter of practicality in the absence of statistical data. Whilst some practitioners test the impact of different distributions as part of their sensitivity analysis, more commonly they tend to rely on the recommendations of the documentation that accompanies the software.

Hall *et al.* (2001) acknowledge the presence of several sophisti-cated risk management packages but argue that the opportunity to communicate risk throughout the supply chain is being lost owing to the lack of a simple tool suitable for newcomers to risk manage-ment. Their research, commissioned by CIRIA, has resulted in the development of a spreadsheet-based software tool, which guides the user through each stage in the process and seems to offer a new and refreshingly straightforward approach. The aim, they state, is to add value by providing a common format for risk communication throughout the supply chain, clarifying risk ownership and providing a convenient and traceable mechanism for revisiting risk assessments as a project proceeds. It will be interesting to observe what level of uptake this latest package achieves over the next couple of years.

Outcomes of the risk management process

Whilst the immediate product of the risk management workshop is commonly a risk register with accompanying action plans and mitigation plans, there appears to be a general consensus that the main outcome of the risk management process is a more realistic estimate of the project budget. Practitioners often use the risk register to assist in building the uncertainties and risks into the cost model. This may take the form of probabilistic estimates for various construction features or elements, but in most cases the overall objective is to produce a more informed assessment of the necessary level of project contingency.

As stated earlier, Monte Carlo simulation is popular, being used to generate cumulative probability profiles and thereby indicate to clients the likelihood of any particular single cost figure being achieved. Commonly cited values of interest to clients are the 50% probability and 80% probability figures. In some instances, for example on PFI projects, this would relate not simply to the capital cost but also to the life cycle costs. At the same time, some practitioners see a danger in forecasting a *most likely* figure that is higher than a previous cost estimate because of the potentially negative reaction by the client. Detailed probabilistic analysis of the programme through the interrogation of critical activity durations, for instance, is less common. Uncertainty regarding time is usually translated into cost and is therefore accommodated within the cost contingency. This perhaps reflects the quantity surveyor background of many risk management practitioners in that they may feel most comfortable when dealing with the project cost parameters.

Evaluation of the service

Evaluation of risk management service provision is variable in practice. At one extreme, no evaluation takes place. Elsewhere, an informal approach following a risk management intervention, for example a workshop, establishes whether the client is satisfied with the delivery. Occasionally this takes the form of a feedback questionnaire, but more often a conversation with the client and individuals within the project team is deemed sufficient. It is acknowledged, however, that such measures are often inadequate and do not really help to improve the service.

The draft risk management report frequently prompts reaction from the client and other members of the team but lack of feedback at later stages in the project can prevent meaningful evaluation. Formal procedures for data collection are rarely in place and therefore much depends upon the willingness of the project team to discuss the issues. At the construction stage, when risks are often passed on to the contractor, the project team may be reluctant to raise the profile of 'nitty-gritty' items, which are seen to be the contractor's responsibility to manage.

Post-project evaluations are sometimes conducted but these tend to embrace the whole spectrum of deliverables and project targets rather than focus purely on risk. The corollary to this is that project archives are seldom used for formative purposes, and improvements in the service come about through the general experience of the practitioners concerned.

Marketing

Marketing risk management services can be problematic. Client scepticism and an apparent unwillingness or inability to recognise the value of risk management can frustrate attempts to actively market the service, especially to new clients. In part this is due to the reluctance of some clients to accept anything other than a single figure estimate. Attempts to create interest through brochures and mailshots are often ineffective and the response to website promotional material is, in the main, negligible. Therefore, the tendency is to promote risk management as an add-on service to existing clients and to target potential clients who are known to use risk or value management techniques. Of course, this approach relies on a heightened awareness of risk management within the clients' own organisation and as a consequence some specialist units organise in-house demonstrations and presentations.

Training

The nature of the risk management service can rely on the managers' quantitative background and their ability to interpret and apply risk management concepts. Education and training play a key role in the development of an organisation's ability to implement rigorous risk management procedures, and companies adopt a variety of approaches. Internal company training sessions and lunchtime CPD seminars are popular, generally of one or two hours' duration. Where specialist units exist, trainers facilitate these sessions, seeking to raise the general awareness of risk management among colleagues. Indeed, transfer of knowledge within organisations often takes place by osmosis through informal meetings and discussions held between colleagues.

Typically, organisations determine training needs as part of an annual review process and run in-house training seminars on specific subjects. However, evidence of employees receiving formal risk management training is limited. On occasion, where the nature of the risk is extraordinary, an organisation might consider bringing in an external consultant. This may occur when the market is particularly unusual and outside the consultant's area of expertise, or if there is a need to undertake complex statistical analysis.

Summary

As might be expected, practice is varied but there is some commonality in certain aspects of the approach to risk management. Many practitioners perceive risk management as a growth area, although the service does not necessarily attract an additional fee. Within the public sector there is a greater awareness, or culture, of risk management but client scepticism remains in the private sector. Risk management interventions are regularly employed at the concept stage, with markedly less involvement throughout the project life cycle. In general, the project cost parameters are used to measure the impact of risks occurring, although programme, specification and business risks are taken into account.

The use of risk management workshops for project teams is prevalent and the production of risk registers is common. There is a reluctance to over-complicate the classification of risks. The use of Monte Carlo simulation through specialist software is widespread as a means of obtaining a greater degree of confidence in project budgets and, particularly, in calculating accurate and realistic levels of contingency funds. Decision trees, spider diagrams and other more complex techniques are rarely, if ever, used. There is limited use of historical data.

In conclusion, current risk management practice could be described as relatively unsophisticated and there is a tendency to take an intuitive approach to assessing risk attitudes. There is a degree of scepticism regarding the usefulness of complex risk analysis techniques and a predisposition to rely on judgement based on experience. Undoubtedly though, in a competitive commercial environment, these are simply pragmatic responses to the amount of time and money clients are willing to invest in the process.

Bibliography

Akintoye, A. S. and MacLeod, M. J. Risk analysis and management in construction. *International Journal of Project Management*, **15**(1) (1997), 31–38.

Amos, J. and Dent, P. Risk analysis and management for major construction projects. *Proceedings of the RICS COBRA Conference*. Portsmouth, 1997.

Association for Project Management. *Project Risk Analysis and Management*. APM, High Wycombe, 1997.

British Standards Institution. *Project Management. Part 3: Guide to the Management of Business Related Project Risk*. BSI, Milton Keynes, 2000, BS 6079-3: 2000.

Edwards, P. J. and Bowen, P. A. Risk and risk management in construction: towards more appropriate research techniques. *Journal of Construction Procurement*, **4**(2) (1998), 103–115.

Ellis, R. C. T. and Wood, G. D. An investigation into risk management services offered by cost consultants on UK construction projects. *Proceedings of the RICS COBRA Conference*. Glasgow, 2001.

Godfrey, P. S. *Control of Risk – A Guide to the Systematic Management of Risk from Construction*. CIRIA, London, 1996, Special Publication 125.

Hall, J. W., Cruickshank, I. C. and Godfrey, P. S. Software-supported risk management for the construction industry. *Proceedings of the Institution of Civil Engineers, Civil Engineering*, **144** (2001), 42–48.

HM Treasury. *Procurement Guidance No. 2 – Value for Money in Construction Procurement*. HMSO, London, 1997.

Institute of Chartered Accountants. *Corporate Guidance for Internal Control*. Institute of Chartered Accountants Internal Control Working Party, London, 1999.

Institution of Civil Engineers and The Faculty and Institute of Actuaries. *Risk Analysis and Management for Projects*. Thomas Telford, London, 1998.

Jackson, S. H., Griffith, A., Stephenson, P. and Smith, J. Risk management tools and techniques used when estimating initial budgets for building projects. *Proceedings of the 13th Annual ARCOM Conference*. Cambridge, 1997, pp. 123–132.

Perry, J. G. and Hayes R. W. Risk and its management in construction projects. *Proceedings of the Institution of Civil Engineers, Part 1*, **78** (1985), 499–521.

Potts, K. and Weston, S. Risk analysis, estimation and management on major construction works. *Proceedings of the CIB International Symposium North Meets South – Commission W92*. Durban, 1996, pp. 522–531.

Royal Institution of Chartered Surveyors. *The Management of Risk: An Information Paper*. RICS Business Services Ltd, London, 2000.

Simister, S. J. Usage and benefits of project risk analysis and management. *International Journal of Project Management*, **12**(1) (1995), 5–8.

Wood, G. D. and Ellis, R. C. T. An exploration of approaches to risk management applications within UK construction projects. *Proceedings of the 17th Annual ARCOM Conference*. Salford, 2001.

CHAPTER FIVE

Classification and control of risk

N. J. Smith

This chapter introduces the concept of classification or categorisation of risk. This is helpful in the identification of risk sources and risk events by making the analyst aware of potential problems for any given project and also in terms of the importance of risk. Human perceptions of risk are often unrealistic, and here again classification systems can help the analyst achieve a more appropriate understanding. The chapter includes some examples of classification systems which might be applied to construction projects.

Control of risk

Risk is not uniform or constant, nor is our perception of it necessarily accurate. Considering the latter first, the probability of a jackpot win on the National Lottery is similar to the chance of being struck by lightning, both about 12 people out of 53 million people per year – but this is not the perception of the typical Lotto player! Consider a large-scale development; almost by definition, it will inherently be riskier than several small-scale developments, yet an increasing number of large developments are undertaken. Why? Because the risk might be short-term, activity-specific and/or easily controllable. Just because there is 'more' risk it may not be more difficult to manage. Hence we need to know how to determine the controllability of risk sources and effects.

Global and elemental risk

An essential aspect of project appraisal is the reduction of risk to a level which is acceptable to the investor. This process starts with a realistic assessment of all uncertainties associated with the data and predictions generated during appraisal. Many of the uncertainties will involve a possible range of outcomes, that is, it could be better or worse than predicted. Risks arise from uncertainty and are generally interpreted as factors which have an adverse effect on the achievement of the project objectives.

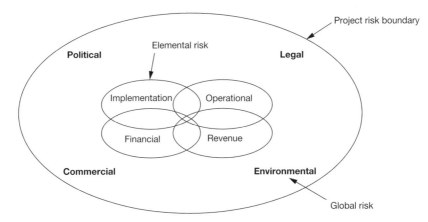

Figure 5.1. Global and elemental risks (Merna and Smith, 1996a)

It is helpful to try to categorise risk associated with projects both as a guide to identification and to facilitate the selection of the most appropriate risk management strategy. One method, proposed by Merna and Smith (1996a), is to separate the more general risks which might influence a project but may be outside the control of the project parties from the risks associated with key project elements; these are referred to as global and elemental risks, and are indicated in Figure 5.1.

Risks will exist beyond the project boundary, but these are usually taken as being too remote and too unlikely to be worth including in the current analysis. The stability of the English legal system for the duration of the project is a risk that could fall into that category. If the English legal system did collapse it would obviously have severe consequences for a project in England, but it is too remote to be actively considered in most project analyses.

Global risks may be capable of being influenced by the principal, if a part of government. Global risks can be subdivided into four sections: political, legal, commercial and environmental risk. Political risk would include events such as a public inquiry, approvals, regulation of competition and exclusivity. Changes in statute law, regulations and directives would all be legal risks. Commercial risks may include the wider aspects of demand and supply, recession and boom, social acceptability, and consumer resistance. Environmental risks are well defined and typical global environmental risks include changes in standards, in external pressure and in environmental consents.

The elemental risks are those risks associated with elements of the project, namely implementation risks and operation risks, and for some projects there will be financial risks and revenue risks. These risks are more likely to be controllable or manageable by project parties. Another way to consider the same information would be to use tables; the examples listed in Tables 5.1– 5.4 were derived by Merna and Smith (1996b).

Table 5.1. Implementation

Risk category	Description
Physical	Natural, pestilence and disease, ground conditions, adverse weather conditions, physical obstructions
Construction	Availability of plant and resources, industrial relations, quality, workmanship, damage, construction period, delay, construction programme, construction techniques, milestones, failure to complete, type of construction contracts, cost of construction, insurances, bonds, access, insolvency
Design	Incomplete design, design life, availability of information, meeting specification and standards, changes in design during construction, competition of design
Technology	New technology, provision for change in existing technology, development costs

Table 5.2. Operational risks

Risk category	Description
Operation	Operating conditions, raw materials, supply, power, distribution of offtake, plant performance, operating plant, interruption to operation due to damage or neglect, consumables, operating methods, resources to operate new and existing facilities, type of operation and maintenance contract, reduced output, guarantees, underestimation of operating costs, licences
Maintenance	Availability of spares, resources, sufficient time for major maintenance, compatibility with associated facilities, warranties
Training	Cost and levels of training, translations, manuals, calibre and availability of personnel, training of principal's personnel after transfer

Table 5.3. Financial risks

Risk category	Description
Interest	Type of rate (fixed, floating or capped), changes in interest rate, existing rates
Payback	Loan period, fixed payments, cash flow milestones, discount rates, rate of return, scheduling of payments, financial engineering
Loan	Type and source of loan, availability of loan, cost of servicing loan, default by lender, standby loan facility, debt/equity ratio, holding period, existing debt, covenants, financial instruments
Equity	Institutional support, take-up of shares, type of equity offered
Dividends	Time and amounts of dividend payments
Currencies	Currencies of loan, ratio of local/base currencies

Table 5.4. Revenue risks

Risk Category	Description
Demand	Accuracy of demand and growth data, ability to meet increase in demand, demand over concession period, demand associated with existing facilities
Toll	Market-led or contract-led revenue, shadow tolls, toll level, currencies of revenue, tariff variation formula, regulated tolls, take and/or make payments
Developments	Changes in revenue streams from developments during concession period

Alternative risk management classification systems

A similar but significantly different system is based upon the ability to control risk (Figure 5.2). The risks are subdivided into 'local', 'global' and 'extreme' in terms of their effect on the project, which is inversely proportional to the ability to control the risks. At the 'local' level, the risks are usually directly under the control of one of the project parties and hence the impact on the project is likely to be small.

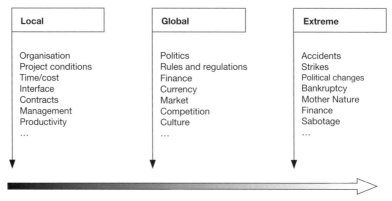

Local	Global	Extreme
Organisation	Politics	Accidents
Project conditions	Rules and regulations	Strikes
Time/cost	Finance	Political changes
Interface	Currency	Bankruptcy
Contracts	Market	Mother Nature
Management	Competition	Finance
Productivity	Culture	Sabotage
…	…	…

Reduced ability to control risk

Figure 5.2. Alternative classification of risk (Smith, 1999)

The 'global'-level risks are subject to a degree of control by one of the project partners, but exactly how much would depend upon the project and the particular parties. Obviously, a government ministry would have more influence over a political issue than would a regional or local authority. The potential impact on the project may be greater but need not always be so.

At the third level of 'extreme', these risks are difficult to control, if they can be controlled at all, by any party to the project. Therefore the impacts will have to be carefully evaluated for each individual project. Sometimes a single uncontrollable risk can be sufficient to cause major problems for the project, the parties or both.

The examples cited at each level are typical but the numbers could be increased by the user to suit the specific types of project being undertaken. Equally, if there is a need to consider programme management whereby several projects are undertaken concurrently, this type of framework is still useful if it assists the parties to identify potential sources of risk. The basic diagram contains flexibility to allow users to derive a more complex classification if required.

Summary

This chapter has shown a number of types of classification system which can be of value to the risk manager. These systems are suitable for customising by experienced users to improve communication, project understanding and the identification of risk sources. There is, of course, no standard system that will guarantee identification of all risks, but classification systems such as these will help to produce better, more complete and more consistent sources of information on which to base the risk management.

Bibliography

Merna, A. and Smith, N. J. *Guide to the Preparation and Evaluation of Build Own Operate Transfer Project Tenders.* Asia Law and Practice, Hong Kong, 1996a.

Merna, A. and Smith, N. J. *Projects Procured by Privately Financed Concession Contracts,* vol. 1. Asia Law and Practice, Hong Kong, 1996b.

Smith, N. J. *Managing Risk in Construction.* Blackwell Science, Oxford, 1999.

CHAPTER SIX
Basic theory of risk management

N. J. Smith

Much of the foundation for the current risk management approaches, including the RAMP methodology discussed in the next chapter, arose from work undertaken at the University of Manchester Institute of Science and Technology (UMIST) in the 1970s and 1980s. Professors Peter Thompson and Stephen Wearne were supported by others, including Professor John Perry, Dr Martin Barnes, Mr Tom Nicholson and Mr Ross Hayes. This chapter contains an overview of the UMIST approach to project risk and reproduces figures that were originally published by the Science and Engineering Research Council (Hayes *et al.*, 1986) and subsequently by Thomas Telford in 1988 (see Thompson and Perry, 1992).

Background

For years the construction industry has had a very poor reputation for coping with risk, with many projects failing to meet deadlines and cost targets. This can be traced to a number of causes, including the inherent difficulties faced when engineers attempt to enact the motto of the Institution of Civil Engineers, and 'harness the forces of nature for the benefits of mankind' (Tredgold, 1828).

Long before formal risk management was recognised as a project function, the likelihood of adverse events delaying things was widely known and two pragmatic approaches for managing the consequences were developed. One approach was to allow some 'slack' in the system, which would be known today as a contingency approach, or to duplicate key resources in a redundancy or premium buying approach. This was largely developed on the basis of custom and practice, with no real understanding of the nature of risk.

In the 1650s Pascal and Fermat laid out the founding principles of the theory of probability, which is the mathematical concept at the heart of the modern concept of risk. By 1730 de Moivre had proposed the structure of a normal distribution, and he was the first to define risk as a chance of loss. Published in 1763, two years after the death of the Reverend Thomas Bayes, Bayes' theorem is the basis for all inference problems using probability theory as logic. Stated as a simple equation, Bayes' theorem shows that

Expected outcome of an event =
Probability of outcome 1 × Value of outcome 1 + ... $P_n \times V_n$

where $P_1 + ... + P_n = 1$.

Bayes' theorem can be applied by hand in decision tree calculations but, increasingly, all types of probabilistic analysis require the use of risk management software packages. The developments in hardware and software over recent years have been such that all but the most sophisticated programs can be accommodated on a PC and an analysis performed in a few minutes. The cost of hardware, software and staff training are relatively small compared with the consequences of the out-turn findings of a risk analysis which might invalidate the economic case for a project, turning a potentially profitable investment into a loss-making failure.

The role of project management

It is suggested that the ultimate burden of responsibility for the identification of risks and their subsequent treatment lies with the owner of the project, the client and/or its project manager. The owner should be motivated to take this responsibility by the threat posed to the achievement of the project objectives – the primary ones being cost, time and quality.

Construction project management has a vital role to play through:

- contributing to sound economic appraisal by producing realistic estimates of cost and time related to an appropriate and defined standard of quality
- achieving efficient project implementation to established targets of cost, time and performance.

In construction projects each of the three primary targets of cost, time and quality is likely to be subject to risk and uncertainty. In terms of project implementation, managers need to be able to undertake or propose actions which reduce or eliminate the effects of risk or uncertainty. In the work of Thompson and Perry the terms 'risk' and 'uncertainty' were both used to represent a threat to or adverse effect on the project performance.

To achieve these aims, it is suggested that a systematic approach to the management of risk is followed:

- identify the risk sources
- quantify their effects (risk analysis)

- develop management responses to risk
- provide for residual risk in the project estimates.

These four stages comprise the process of risk management.

The aims of risk management are, firstly, to ensure that only projects which are genuinely worthwhile are approved and, secondly, to avoid excessive overruns which can invalidate the economic case for the project, usually bring embarrassment and tend to discredit the parties concerned. Risk management can be one of the most creative tasks of project management. It achieves this through generating realism and thereby increasing commitment to control, and through encouraging problem solving, opening the way to innovative solutions to project implementation.

Risk identification

It is generally the case that the realism of the estimates increases as the project proceeds. However, the major decisions are made early in the project life – at appraisal and sanction. It therefore follows that all potential risks and uncertainties which can affect performance, and act as constraints on the project, should be identified early in the project's life.

There is a second, but equally important, reason for the early identification of risks and uncertainties, namely that it focuses project management attention on policies and strategies for the control and allocation of risk. In particular, it highlights those areas where further design, development work, investigation or clarification is most needed.

Risk identification can be of considerable benefit, even if no other stages of risk management are undertaken. The constructive nature of this exercise is worth emphasising again, since it is in direct contrast to the notion, sometimes expressed in industry, that attention to risk creates undue pessimism.

Identification is the first and most significant phase of the risk management process. It brings considerable benefits in terms of project understanding and provides an early indication of the need for risk management strategies. It is not possible to know if all possible sources of risk are ever identified but it is likely that there will be some unknown unknowns. The purpose of the identification process is to use a combination of methods to try to ensure that the extent of the unknown unknowns is as small as is possible.

The identification list is usually compiled from three main techniques, although other possibilities may exist for particular projects:

Box 6.1. *Major categories of sources of risk*

- Client/government/regulatory agencies
- Funding/fiscal
- Definition of project
- Project organisation
- Design
- Local conditions
- Permanent plant supply
- Construction contractors
- Construction materials
- Construction labour
- Construction plant
- Logistics
- Estimating data
- Inflation
- Exchange rates
- *Force majeure*

Box 6.2. *Discrete sources of risk: design*

- Adequacy to meet need
- Experience/competence of design organisation
- Effectiveness of coordination:
  - between design offices
  - between contractor's design and client/consultant design
- Degree of novelty
- Appropriateness of design:
  - to logistics and access
  - to climate
  - to maintainability
  - to operability
- Realism of demands on construction
- Soundness of design data
- Realism of design programme:
  - in relation to design resources
  - in relation to construction needs
  - in relation to special requirements
- Likelihood of design change

- checklists from industries/organisations with relevant experience
- brainstorming with key project participants and other project stakeholders
- historical precedence from 'similar projects' – if any.

An example of a professional checklist of major categories of sources of risk is given in Box 6.1. An example for the design risk category is shown in Box 6.2. It is suggested that it would be useful for project managers to develop their own detailed lists appropriate to the type of projects with which they are usually concerned.

The output from this process might be an extremely long list; possibly several hundred potential risk sources might have been identified. Ideally, a much smaller listing of 5–15 major risks would be more practicable. Therefore the risks are assessed on two criteria, their impact on the project and their likelihood of occurrence. Only those risk sources with both a high impact and a high probability should be retained. It is possible that the risk management process could terminate at this point but it is more likely that some assessment of the effects of these risk sources will be required.

Risk analysis

The purpose of risk identification is to quantify the effects on the project of the risks identified. The first step is to decide which analytical technique to use. At the simplest level, each risk may be treated independently of all others with no attempt made to quantify any probability of occurrence. Greater sophistication can be achieved by incorporating probabilities and interdependence of risks into the calculations but the techniques then become more complex. The choice of technique will usually be constrained by the available experience, expertise and computer software.

Whichever technique is chosen, the next step requires that judgements are made of the impact of each risk and, in some cases, of the probability of occurrence of each risk and of various possible outcomes of the risk.

The approach proposed below suggests three tiers of risk analysis, as follows.

Elementary risk analysis

At the crudest level, the individual major risk categories can be aggregated into two or three major risk effects and a subjective but experienced judgement made of their effect on cost and time. For example, a minimum rate of return in excess of minimum bank lending rate could be considered. This may be adequate when comparing alternatives at the appraisal stage.

Sensitivity analysis

The basis of a sensitivity analysis is to define a likely range of variation for elements of the project data. The final project cost or duration is then assessed for each variation in the data. In effect, a series of 'what if' estimates is produced.

The results of sensitivity analyses are often presented graphically, which readily indicates the most sensitive or critical areas for

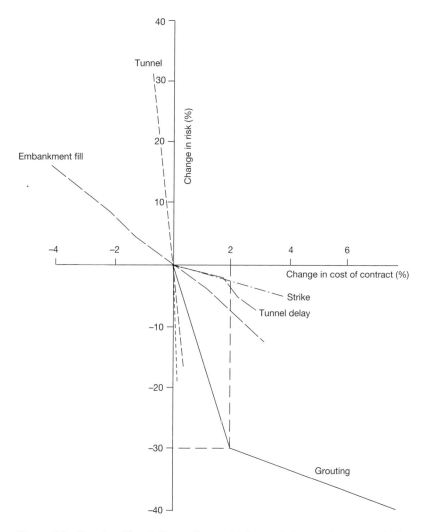

Figure 6.1. Dearden Clough Reservoir: sensitivity analysis covering uncertainties in estimated construction cost. A 30% decrease in output when grouting the foundations adds 2% to the total cost of the contract; any greater reduction is far more serious

management to direct its attention towards. Figure 6.1 illustrates a typical sensitivity diagram, or spider diagram, for a reservoir project.

One weakness of sensitivity analysis is that the risks are treated individually and independently. Caution must therefore be exercised when using the data directly to assess the effects of combinations of risks.

Probability analysis

Probability analysis is a more sophisticated form of risk analysis. It overcomes the weaknesses of sensitivity analysis by specifying a probability distribution for each risk and then considering the risks in combination.

The 'Monte Carlo' approach is used to sample from the total number of combinations of risks. The steps are:

(1) Assess the ranges of variation for the uncertain data and determine the probability distribution most suited to each piece of data.
(2) Randomly select values for the data within the specified range, taking into account the probability of occurrence.
(3) Run an analysis to determine values for the evaluation criteria for the combination of values selected.

Repeat steps (1) and (2) a number of times. The resulting collection of outcomes is arranged in sorted order to form probability distributions of the evaluation criteria. The accuracy of the final distribution depends on the number of repetitions, or iterations, usually between 100 and 1000.

Since the outcome from the Monte Carlo analysis is a collection of, say, 1000 values for each evaluation criterion, it is unlikely that the same value for the evaluation criterion will be calculated more than a small number of times. The values are therefore grouped into class intervals. The results are presented as frequency and cumulative frequency distributions; see Figures 6.2 and 6.3.

Risk response

Risk response can be considered in terms of avoidance or reduction, transfer, or retention.

Avoidance or reduction

In the extreme, risks may have such serious consequences as to warrant a reappraisal of the project or even the replacement of the project by an alternative project.

It is perhaps more likely that risk identification and analysis will indicate the need for redesign, more detailed design, further site investigation, different packaging of the work content, alternative contract strategies or different methods of construction in order to reduce or avoid risk.

Transfer

The four most common routes for the transfer of risk in construction projects and contracts are:

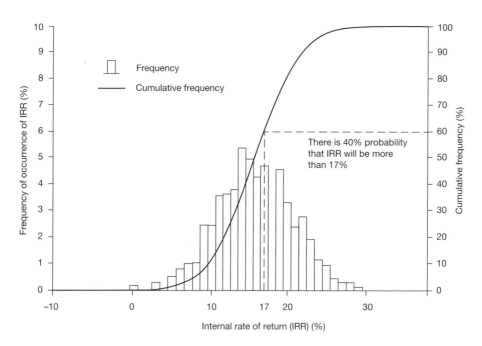

Figure 6.2. Mining venture frequency diagram

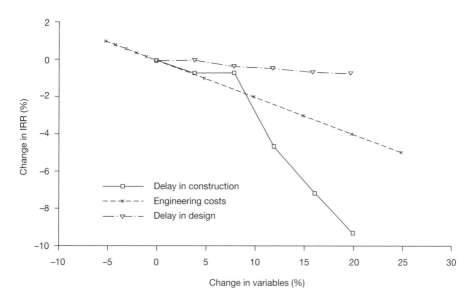

Figure 6.3. Design and construction risks: sensitivity diagram

- client to contractor or designer
- contractor to subcontractor
- client, contractor, subcontractor or designer to insurer
- contractor or subcontractor to surety.

The essential characteristic of the transfer response is that the consequences of the risks, if they occur, are shared with or totally carried by a party other than the client. The client should expect to pay a premium for this privilege. The responsibility for initiating this form of risk response therefore lies with the client, and the client should ensure that it is in its own best interest to transfer the risk. This will necessitate consideration of both the client's own and the other parties' objectives, the relative abilities of the parties to assume the risk, the degree of control over the situation, the potential gain or loss, and the incentive.

Retention

Risks which are retained by either party to a contract may be controllable or uncontrollable by that party. Where control is possible, it may be exerted to reduce the likelihood of occurrence of a risk event and also to minimise the impact if the event occurs.

Summary

Risk management should be viewed constructively and creatively. Rigid application of a set technique or procedure is not advocated or encouraged. Indeed, methodologies are, relatively speaking, in their infancy and are evolving with practice. The initiative for the application of risk management rests with clients and their professional advisers, particularly the project manager.

Bibliography

Hayes, R. W., Perry, J. G., Thompson, P. A. and Willmer, G. *Risk Management in Engineering Construction*. Science and Engineering Research Council, Swindon, 1986, Project Report.

Thompson, P. A. and Perry, J. G. (eds). *Engineering Construction Risks*. Thomas Telford, London, 1992.

Tredgold, T. Charter of the Institution of Civil Engineers. London, 1828.

CHAPTER SEVEN

Risk analysis and management of projects: RAMP

N. J. Smith

The RAMP approach has been developed jointly by the Institution of Civil Engineers and the actuarial profession. The approach offers a strategic framework for managing project risk and its financial implications. The full RAMP guide (Institution of Civil Engineers *et al.*, 2002) is published by Thomas Telford in its own right, and the author would like to acknowledge the cooperation of the RAMP Working Party in the preparation of this chapter and for permission to reproduce small sections and a figure. However, the purpose of this chapter is to give an insight into the basic components of the RAMP approach.

RAMP principles

In 1994 a joint working party of professional actuaries, civil engineers and economists was established to develop a better way of considering project risk and controlling it effectively. The synergy of the joint approach, which was first published in 1998, resulted in an advance in the management of risk in capital investment projects.

The RAMP approach was designed to be rigorous and repeatable and has been recommended by HM Treasury in its 'Green Book' outlining project appraisal and evaluation techniques. RAMP is a comprehensive and systematic process which can be conducted at a strategic level or as a detailed project analytical and control process, as required.

The RAMP process is based on four key activities:

- *Process launch*: definition of project objectives, scope and initial plans, together with essential underlying assumptions.
- *Risk review*: an iterative activity occurring at decision points throughout the project life cycle, when risks are identified and entered into a form of risk register. The impact of the risk and its likelihood of occurrence are considered and, where

appropriate, risk mitigation measures are proposed. For the remaining risks, a risk response plan is prepared.

- *Risk management*: is conducted between risk reviews and involves the implementation of both risk mitigation and risk response plans.
- *Process close-down*: retrospective review of the investment in terms of meeting the project objectives.

Prerequisites for RAMP

In order to utilise the RAMP process, it is necessary to have a good understanding of risk and investment and the input needs of the process. In the RAMP system, 'risk is defined as a threat (or opportunity) which could affect adversely (or favourably) achievement of the objectives of an investment.' This is slightly different from the approach taken earlier in this book, where 'risk' and 'uncertainty' have been defined separately as known and unknown, which could be either adverse or favourable.

The possibility of unknown unknowns has to be considered. It is never possible to be certain how significant these may be, but the purpose of RAMP is to improve project decision-making and not to try to predict an actual outcome. Nevertheless, one of the key aspects of the RAMP process is to strive to identify as many of these potential unknowns as is possible.

A basic understanding of likelihood or probability is required. For each risk source identified, it should be possible to evaluate the impact of the risk, usually in financial terms. Somewhat more difficult is the assessment of the likelihood of the risk occurring. Some risks are easy to assess and others much more difficult, and it may be that the probability can only be expressed as a range, for example 0.10–0.25 probability. Further investigation might be necessary to narrow the range if the risk has significant adverse consequences. For each risk, the likelihood multiplied by the impact gives a measure of the value of the risk, usually known as the expected value. This allows the ranking of risk to be undertaken and provides an indication of how much money might be invested in risk mitigation.

Investment life cycle

RAMP has six separate but sometimes overlapping stages, where it is used in conjunction with:

- *Opportunity identification*: the project investment is identified and an initial decision on whether to proceed to a full appraisal is made.

- *Appraisal*: determination of objectives, scope and requirements; business case; and feasibility study as to whether to invest or not.
- *Investment planning*: procuring of funding, preliminary design, moving towards financial closure.
- *Asset creation*: the project – planning, designing, procuring, constructing and commissioning the asset.
- *Operation*: utilising and managing the asset, including maintenance and training.
- *Close-down*: end of the investment cycle, the asset is either sold or decommissioned.

Table 7.1 shows the main objectives, activities, associated parameters and risk management processes in each stage of the investment life cycle. A way of financially evaluating the risks to which an investment is exposed is required and this is most commonly satisfied by some form of cost model, either cash flow based using a spreadsheet or activity based using network analysis methods. The model should also be able to cope with expressing values using discounted parameters, in particular net present value (NPV) or internal rate of return (IRR).

RAMP processes

The four basic processes can be expanded to identify the main action points required to execute a RAMP analysis. Figure 7.1 uses a flow diagram format to illustrate the main features of the process.

Most of the procedures are well known or straightforward, but there are some aspects which can be problematic and deserve further consideration. One of these aspects occurs in Activity B, risk review, and concerns the measures for mitigating risk.

Mitigation is a fundamental part of the process but is not always well understood. Briefly, risk mitigation is concerned with the lessening of the adverse impacts of risk, and its effectiveness is a key requirement for the successful management of risk. There are four main approaches to risk mitigation, namely reduction, transfer, avoidance and absorption; any remaining risk is known as the residual risk. Reduction, transfer and avoidance are well understood but there can be confusion over risk absorption.

Risk absorption can occur when two or more parties to a project are each able to exercise partial control over the incidence and/or the impact of a risk and can reach agreement on the equitable sharing of any adverse consequences. The sharing or pooling of risk amongst joint ventures, partnerships or consortia is becoming more common with the increased incidence of these forms of procurement in construction. Once absorbed, the risk is removed from the residual risks.

Table 7.1. Activities, key parameters and RAMP process in each stage of investment life cycle

Investment stage/ objective	Principal activities	Key parameters	RAMP process
Opportunity identification To identify opportunity and decide whether it is worthwhile conducting a full appraisal	Identify business need Define investment opportunity Make initial assessment Decide whether to proceed with appraisal	Broad estimate of capital cost and cash flows Cost appraisal	Preliminary review
Appraisal To decide whether the investment should be made	Define investment objectives, scope and requirements Define project structure and strategy Develop business case Identify funding options Conduct feasibility study Decide (in principle) whether to proceed with investment	Refined estimates of capital cost and cash flows Cost of investment planning phase	Full risk review
Investment planning To prepare for effective implementation of the project	Procurement of funding Obtaining planning consents Preliminary design work Compiling project implementation plan Placing advance contracts (e.g. site preparation) Making final decision to proceed with investment	Financing cost Refined estimates of capital cost and cash flows	Risk review (prior to final decision)
Asset creation To design, construct and commission the asset, and prepare for operation	Mobilising the project team Detailed planning and design Procurement/ tendering Construction Testing, commissioning and handover Ensuring safety Preparing for operation	Project objectives: • scope[a] • performance/ quality[a] • timing[a] • capital cost	Risk reviews (during or towards end of each activity) and risk management between risk reviews

Table 7.1. Contd

Investment stage/ objective	Principal activities	Key parameters	RAMP process
Operation To operate the asset to obtain optimum benefits for client and other principal stakeholders (including investors and customers)	Operating the service Deriving revenue and other benefits Maintaining and reviewing the asset	Operating cost Maintenance cost Cost of renewals Revenue Non-revenue benefits	Risk reviews (periodically)
Close-down To complete investment, dispose of asset and related business, and review its success	Sale, transfer, decommissioning or termination of asset and related business Post-investment review	Decommissioning cost Cost of staff redundancies Disposal cost Resale or residual value	Final risk review and RAMP close-down

ªThese have a potential impact on one or more financial parameters

Using RAMP at the appraisal stage

At the early stages of a project, there is often a need for a simple process to evaluate competing project options to determine viability and to obtain, for the viable options, an indication of priority. The RAMP methodology can be used if combined with a suitable financial investment model.

Initially, the objectives of the project must be clearly established. This is an important stage, and unless the purpose of the project is understood it is unlikely that decisions will be made in the most effective manner. Following on from the objectives, an assessment of the scope and requirements of the project should be undertaken to establish the extent of practicable safety measures and possible future extensions or changes of use of the facilities, and all this data should be incorporated into a project plan.

Once a time schedule has been established a forecast of the cash flows should be made. Where it is possible, the cash flow forecast should be checked against similar projects. The usual convention is that cash expenditures are negative and cash revenues are positive. The cash flows are typically based on annual values, with discounting calculations used to assess the time value of money in terms of the NPV. The NPV initial value assumes that no risks occur.

The identification and analysis of major risks in the project is now required. These events can be recorded in a risk register, where, for

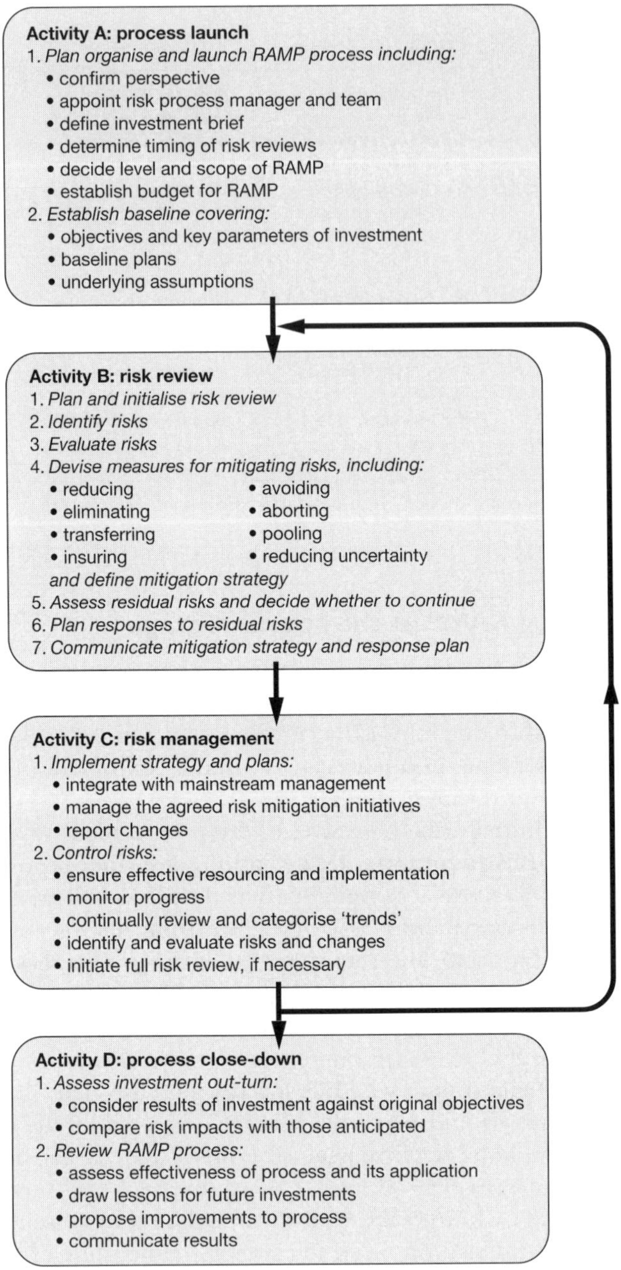

Activity A: process launch
1. *Plan organise and launch RAMP process including:*
   - confirm perspective
   - appoint risk process manager and team
   - define investment brief
   - determine timing of risk reviews
   - decide level and scope of RAMP
   - establish budget for RAMP
2. *Establish baseline covering:*
   - objectives and key parameters of investment
   - baseline plans
   - underlying assumptions

Activity B: risk review
1. *Plan and initialise risk review*
2. *Identify risks*
3. *Evaluate risks*
4. *Devise measures for mitigating risks, including:*
   - reducing
   - eliminating
   - transferring
   - insuring
   - avoiding
   - aborting
   - pooling
   - reducing uncertainty
   and define mitigation strategy
5. *Assess residual risks and decide whether to continue*
6. *Plan responses to residual risks*
7. *Communicate mitigation strategy and response plan*

Activity C: risk management
1. *Implement strategy and plans:*
   - integrate with mainstream management
   - manage the agreed risk mitigation initiatives
   - report changes
2. *Control risks:*
   - ensure effective resourcing and implementation
   - monitor progress
   - continually review and categorise 'trends'
   - identify and evaluate risks and changes
   - initiate full risk review, if necessary

Activity D: process close-down
1. *Assess investment out-turn:*
   - consider results of investment against original objectives
   - compare risk impacts with those anticipated
2. *Review RAMP process:*
   - assess effectiveness of process and its application
   - draw lessons for future investments
   - propose improvements to process
   - communicate results

Figure 7.1. The RAMP process

each risk scenario, the upside or, more frequently, downside event is recorded. An estimated probability of occurrence is given, with a valuation of the expected impact on the project. The cash flow model is then rerun using each of the principal risk scenarios, showing the effect on the project NPV.

The risks may be suitable for mitigation, and a careful study and brainstorming of possible risk mitigation is a crucial stage in the RAMP process. The scenarios can be reassessed taking into account realistic mitigation measures to see whether the resulting NPV risk profile is likely to be more attractive to a client who would find losses hard to bear.

The process depends upon access to good-quality data and a robust cash flow model. It also requires the user to exercise judgement and creativity in the identification of risk and in the types of risk mitigation. It is important to try to remove bias in the appraisal despite the natural pressures to want projects to be approved.

RAMP on a major project

Risk is a dynamic process, and in the application of RAMP to a major project this will involve an iterative approach providing more detail and more certainty as the project moves along the project life cycle.

Initially, a preliminary appraisal would be undertaken as described in the section above.

A secondary risk assessment would then be undertaken. This would involve brainstorming of the project team and other associated professionals, specialist, historical data if available, and possibly discrete risk studies into particularly complex elements, for example wind and tide conditions for marine works. At regular intervals reviews may be performed with a view to generating modifications which may reduce risk or improve financial return. Comparison with the client's strategic time and budget allowances will ascertain whether it is worth proceeding to the later stages of the analysis.

At this stage it is usual to construct a mathematical model of the project. This is most often executed utilising commercial software packages. Although there are many different modelling methods, the two most common mathematical project models are spreadsheet cash flow models and critical-path network activity models.

The model is then used for testing the 'what if' scenarios for risk mitigation or opportunity exploitation. The possible combination of outcomes can be evaluated on the model and a preferred option identified.

The new option can now be investigated in greater detail. The third assessment would identify additional risk sources arising from the project decisions and choices already made. Once again, each of the additional risk areas is considered in detail and the possibilities for risk avoidance or mitigation are examined. The model is then rerun to consider the revised circumstances and to check if the option is still the best project choice. If not, then the process is repeated until a consistent scope for a preferred project is established.

A formal RAMP report can now be drafted, based upon the findings of the model analysis. The report should contain:

- a listing of the key risks and their respective mitigation plans
- a discounted cash flow calculation – enhanced if possible by an indication of the optimistic and pessimistic range of outcome values
- feedback mechanisms as part of a control cycle
- a recommendation for the decision makers on whether to proceed and, if so, the preferred project option.

The RAMP report may also have a role in informing project investors and financiers and increasing confidence in the analysis on the basis of:

- a full listing of key risks and their evaluation
- a range of risk mitigation methods
- clearly identified and quantified residual risks.

The report will form the basis for the decision to sanction the project. The decision makers will also use their own experience and expertise and may on occasion require an independent view on certain aspects of a project. The decision will sanction the project, determine the preferred option and approve the appropriate project management team.

RAMP implementation and control

The RAMP report is then used to manage risks as part of the overall project management strategy. Naturally, the identification, analysis, mitigation and plans need to be monitored and updated regularly throughout the implementation phase. These two processes need to be fully integrated.

Monitoring will identify significant differences between plan and implementation, but sometimes this may be due to a poor plan. It is therefore necessary to investigate differences and to establish the reasons for the divergence before taking a management decision. The RAMP guide recommends monitoring trends, which can be identified through:

Table 7.2. Key documents created in the RAMP process

Document	Purpose	Contents
RAMP process plan	To define strategy and basis for undertaking RAMP process over whole life of investment	Investment brief and perspective, organisation and strategy for RAMP process and baseline information
Risk diary	To record significant events, issues and outcomes during RAMP process	Significant events, problems, results, ideas for improvement and unforeseen risks arising
Risk review plan	To describe the plan for carrying out a specific, individual risk review	Risk process manager and review team Purpose, scope and level of review Action plan, resource requirements, budget and timetable
Risk register	To record risk events and analyses	Risk schedules: • preliminary list of risks • refined list of risks • groups of risks • mitigated risks • residual risks Individual risk analyses Risk diagrams Assumptions list
Risk mitigation strategy	To define the measures to be adopted to avoid, reduce or transfer risk	Mitigated risks, mitigation measures, costs of mitigation and secondary risks
Risk response plan	To define plans for containing or responding to residual risks	Containment and contingency plans and associated budgets. Responsibilities for action
Investment model runs	To record the data and results of each run of the investment model	Timing and purpose of run Scenarios modelled Parameter values Resulting NPVs
Risk review report	To summarise and report on results of risk review	Main risks and potential effects Summary of plans Riskiness of investment Lessons learnt Significant changes arising from review
Trend schedules	To identify, evaluate and act on new risks or changes in risk exposures and outcomes during the ongoing management of the investment	Events, situations and changes (trends) which could affect risks, categorised into • potential • expected • committed
RAMP close-down report	To report on overall performance of investment and effectiveness of RAMP process	Comparison of investment plan (as authorised) against out-turn result Summary of risk history Assessment of RAMP process as applied to investment Suggested improvements to RAMP process for future use

- site visits
- progress reviews
- design meetings
- correspondence
- negotiations
- ground surveys
- market research
- tests
- reports on similar investments.

Throughout the implementation phase, the fundamental merits of the project are assessed and reassessed to ensure that informed decision-making can be performed at all times.

Closing-down phase

At the end of the implementation phase, RAMP can be used to evaluate the performance by comparing anticipated and actual risk and by comparing the actual out-turn with the original objectives. The project manager will critically assess the effectiveness of the process, and the results will be recorded in a RAMP close-down report.

The RAMP process generates a series of documents, and these are listed in Table 7.2.

Bibliography

Institution of Civil Engineers and the Faculty and Institute of Actuaries. *Risk Analysis and Management for Projects*, revised edition. Thomas Telford, London, 2002.

CHAPTER EIGHT
Uncertainty management

P. W. Hetland

During the 1990s, the development of a body of knowledge of uncertainty management took place. The European Institute of Advanced Project and Contract Management (Epci), in close cooperation with risk management experts drawn from leading European universities in the field of project management, made a major contribution to this development. This chapter is based on developments achieved by Epci that are not yet known to a broader audience of project management professionals.

In general, a major shift in the perception of risk management is currently taking place. Risk is not considered something purely adverse any more. There are upside potentials as well as downside risks; hence the name 'risk management' is being replaced by uncertainty management or even, by some, management of uncertainty and risk, as discussed by Chapman and Ward (2002). This chapter is about understanding and managing project uncertainties and their implications. The author has deliberately chosen the term 'uncertainty' to clarify the assertions that:

- risk is an implication of a phenomenon being uncertain
- implications of an uncertain phenomenon may be wanted (upside potentials) or unwanted (downside risks)
- uncertainties and their implications need to be understood properly to be managed successfully.

The chapter is subdivided into three major themes:

- major trends and paradoxes in managing construction projects
- what is uncertain and why
- managing project uncertainties and their implications.

The chapter is deliberately biased, as exploring the potential value of project uncertainty is in focus at the expense of the traditional linear, reductionistic approaches to uncertainty and risk management catered for elsewhere in this book. Consequently, the chapter will cite many publications.

Major trends and paradoxes in managing construction projects

A first approach to a comprehensive understanding of the value of the upside potentials is a survey and interpretation of recent trends and emerging paradoxes at the cutting edge of the project management profession. The following paragraphs give a brief account of experiences from onshore and offshore projects in Europe over recent decades. In particular, developments originating from the CRINE (1994) and Norsok (1995) initiatives are discussed. The identified trends of core interest may be grouped into four major themes:

- life cycle orientation
- functional requirements
- partnering
- differentiation.

A number of full-scale experiments on real projects under the CRINE and Norsok regimes have demonstrated substantial savings or improvements in project costs and durations. The construction industry is, however, reluctant to explore the benefit of strategies directed towards continuous enhancement of project value. The same is true for owner companies of major capital projects. Surprisingly, there has been little interest from onshore industries in studying the 'quantum leap' made by offshore industries, with the exception of the ACTIVE initiative in the UK (Hetland, 2001a).

A threshold of significant importance has been identified, which has to be understood and overcome in order to harvest the upside potentials of major construction projects. The threshold is described in terms of the prevailing paradoxes: the Juran dilemma, the Kaasen paradox and the Lofthus syndrome. In summary, these paradoxes say that higher project value requires acceptance of higher project uncertainties, which result in unpredictable project outcomes. The environments of media-sensitive construction projects require predictable project results, which contradict the requirement for continuous improvement of project value.

Life cycle orientation

Traditionally, construction projects are structured in a number of sequential phases. The core phase, the execution phase, often represents a *closed system*. Everything is defined in the previous engineering phase. Most national and international standard contracts for construction work assume that detailed drawings and lists of materials exist prior to contract award, hence eliminating the freedom of contractors to interact with engineers in the design process.

There are exceptions to the standard construction, build-only contracts. Design–build contracts are known to most practitioners, but are not widely used in onshore construction arenas. In the oil and gas industries, design–build contracts (otherwise referred to as EPC or Epci contracts) were used extensively in the 1980s and 1990s. Other variations to the standard contract from a life cycle perspective are concepts such as turnkey and concession contracts.

Under the regimes of CRINE and Norsok, the traditional construction project is closely linked to the underlying business case. The project is no longer reduced to assemblies and the outfitting of a physical object, say a building or an offshore production facility. The focus is on what the object is for. Owners do not buy project objects any more – they buy a service for satisfaction of their needs.

Obviously, cost considerations are extended from construction costs to life cycle costs and life cycle values. However, the new perspective is not limited to summing all costs and values; the main focus of the new wave is to engineer project costs. Value engineering, value management and financial engineering are examples of new value-enhancing activities. In quality terms, we see a shift of interest from 'according to specifications' to 'fitness for use', from detailed specifications to functional requirements.

Functional requirements

The traditional prescriptive contractual specifications impose extensive requirements on the manufacturer or contractor, such as changes to their working practices, excessive documentation and unreasonable, onerous regimes of traceability, certification and inspection. These prescriptive requirements from project owners and engineering contractors are often preferential, and are given precedence over supplier standards. Such overspecification can be further hindered by numerous cross-references to industry codes and standards with a lack of guidance on conflicts and the extent of applicability. This stifles supplier initiatives and inhibits the use of standardised equipment.

The CRINE initiative suggested a shift in roles between owner and contractor (purchaser and supplier). Prescriptive specifications are replaced by functional requirements. The ultimate goal, according to CRINE, for purchase specification is where the purchaser limits input solely to the functional requirements. This provides the supplier with the maximum amount of flexibility to provide a cost-effective bid based around standard equipment. This ideal state of affairs requires (CRINE, 1994)

Commitment by all purchasers (oil companies and contractors) to work to an agreed functional specification content.

Suppliers grasping opportunities to provide standard equipment and undertake increased management and self-checking responsibilities.

Development of international standards and codes, embodying good engineering practice and having common currency in the industry.

In summary, the CRINE suggestion to focus on function makes the contractor accountable for performance reinforced through an incentive to improve, resulting in the use of standard equipment, shorter delivery and reduced costs.

Partnering

Both CRINE and Norsok postulate that cooperative strategies are superior to competitive strategies on offshore development projects. Firstly, cooperative strategies are better than competitive ones simply because no effort is wasted on adversarial/opportunistic attitudes/ actions. Secondly, cooperation opens up a challenging dialogue between complementary project stakeholders directed towards continuous value enhancement over the whole project lifetime. Thirdly, cooperative strategies make it possible to employ suppliers at the earliest project stages, prior to vital project decisions being made.

A prerequisite for successful partnering is, however, that the part-nering parties spend time on understanding the value functions of participating organisations, align their project goals and develop a project culture characterised by a high level of trust and mutual respect.

Differentiation

Projects are different, strategies are different and project environ-ments are different (Hetland, 2001b). There is no such thing as a universal best practice, or best theory for that matter. Projects are unique, environments are unique and expectations of project delivery vary with the level of ambition of the core stakeholders. Project goals and strategies are chosen to meet the appropriate level of ambition given numerous situational constraints.

According to Obeng, projects may be classified as closed, semi-closed, semi-open or open (Obeng, 1999). Closed projects are char-acterised by the fact that you know what and know how, open projects by the fact that you don't know what or how. On semi-closed projects you know what but not how, and on semi-open projects you know how but not what.

The Obeng space can be operationalised and used to combine the project challenge with the selection of an appropriate execution strategy (Hetland, 2000). According to this approach, closed projects are characterised by detailed definitions of project work scope and prescriptive procedures for executing project work. Semi-closed projects are characterised through functional requirements and full freedom to the supplier as to how to achieve these requirements. Another way of describing a semi-closed project is to set stretched targets, say to cut project cost or duration by 25% compared with a given benchmark based on a similar project of the past. Semi-open projects are characterised by guide documents directed towards the project work processes. CRINE and Norsok are examples of such guidance.

Epci has developed a number of generically different project strategies, grouped into four major areas according to purpose, roughly related to the Obeng categories. Group A strategies are, in particular, appropriate to cope with closed projects, and focus on copying best practices and extensive control such that actual project processes run according to detailed predefined procedures. Group B and C strategies are designed to cope with semi-closed and semi-open projects. Group D strategies are directed towards open projects.

The strategy processes are different in nature. A strategies are linear; B, C and D strategies are complex. A strategies are linear-reductionistic and focus on a stepwise definition of project work and reduction in project uncertainties. Traditionally, linear and linear-reductionistic processes have been applied to construction projects. Management of these kinds of processes is what the PMI PMP certification of project managers is all about.

The ambitions to bring down project cost by 40% by application of Norsok principles (in a semi-closed or semi-open project challenge) require competencies in managing 'Bwh' (Bwh is a subset of group B, where w states that a stretched target is given and h that a certain guidance philosophy shall be applied). This requires knowledge and experience that currently are unknown to most practising project managers. Though the new project challenges differ significantly from those of past projects, the practitioners continue to carry out project work as before.

Review of project types and strategies

The Juran dilemma

The famous guru on quality management Dr Joseph Juran once said, 'If you always do what you always did, you will always get what you always got'.

The statement may be interpreted as meaning that if you want to achieve the same result once more, you simply have to repeat what you did last time. Another interpretation of the same statement is that if you want to achieve something different, say an improvement compared with what you used to get, you either have to do the same thing differently or do different things. The consequence of the latter interpretation is that any improvement requires some kind of change in or of work processes – improvements imply escaping from the Juran dilemma.

Construction of facilities for exploiting oil and gas from the Norwegian continental shelf commenced in the early 1970s. The first-generation projects overran budgets by 176% on average (Olje- og energidepartementet, 1980). The second-generation projects also overran their budgets but by lesser amounts. The third-, fourth- and fifth-generation projects were not significantly better controlled, but budgets became more realistic as as-built data became available. The Juran statement made sense, as it was by now apparent that repetition ensured predictability. As the companies involved now always did what they used to do, they always got what they used to get, and consequently there were no major surprises concerning delays or budget overruns any more.

In the early 1990s the project environments changed dramatically. The price of crude oil dropped to half of what it used to be. Long-term predictions of crude oil prices were not optimistic at all. In order to regain profitable investments in the North Sea, project costs had to be cut significantly. Cost reductions of 40% were suggested by oil companies operating on the Norwegian shelf. To achieve this kind of an improvement industry had to escape the Juran dilemma; it had to do projects differently to achieve a 40% cost saving.

The Kaasen paradox

A recipe for escaping the Juran dilemma in the UK was the CRINE initiative (later extended to onshore construction through the ACTIVE initiative). A similar recipe was developed in Norway, Norsok. Through these initiatives, a number of changes to project work processes were suggested:

- change of business culture
- redefining of roles between owner and suppliers
- alignment of owner and supplier goals
- functional requirements
- incentivised contracts.

In summary, the CRINE and Norsok initiatives introduced a variety of previously unknown uncertainties. Project challenges were typically semi-closed or semi-open by nature, and the strategies applied B*wh* instead of the traditional BA strategy. Consequently, project managers had to manage uncertainty rather than reduce uncertainty. Uncertainty management became one of the new competence requirements of project management. Rather than viewing uncertainty through the well-known models of risk analysis, focusing on what might go wrong and why, the focus of the new wave was on 'exploring the value of uncertainty – turning risks into value and harvest upside potentials' (Chapman, 1998).

This major change of perspective caused some quite unexpected outcomes, ranging from extraordinary results to major financial upheavals. Generally, application of CRINE was promising, while Norsok was considered somewhat disappointing (Østensen, 1999).

In Norway, some larger projects overran cost budgets by substantial amounts of money, causing a governmental enquiry, the Kaasen study (Olje- og energidepartementet, 1999). Kaasen, a professor of law at Oslo University, concluded that a 27% budget overrun was a 13% cost saving – later called the Kaasen paradox.

The meaning of this obviously contradictory statement is easy to grasp if you understand how estimates and budgets were prepared under the Norsok regime. Firstly, project cost estimates were sanctioned by the project owner at an earlier stage compared with pre-Norsok projects. Therefore the uncertainties were higher. Secondly, the estimates were based on pre-Norsok data. The resulting expected values of the uncertain estimates, representing a benchmark of pre-Norsok best practice, were then stretched by 40%. Practically, this means that the project owner expects to cut costs by 40% when applying the Norsok principles.

The Kaasen study highlighted the fact that for some major projects the stretched target of 40% was not achieved. The actual improvement was only 13%; the outstanding 27% was then interpreted by major stakeholders as a budget overrun of a substantial size. It was understood neither that projects under Norsok were run at a far higher level of uncertainty than previously experienced, nor that the new project budget represented a target stretched by an almost unbelievable 40%. The biased misconception of the new way of managing projects led to the fall of the board of directors and the top management team of a major Norwegian oil company. The fact that the budget overrun simultaneously represented a major improvement over best practice compared with the pre-Norsok benchmark is still not generally acknowledged, neither by project professionals nor by non-professional stakeholders of major projects.

The Lofthus syndrome

Mr M. Lofthus, a former CEO of Tate and Lyle, once expressed 'regret risk' phenomena the following way: 'If you first publish a number, it will haunt you for ever' (Lofthus, 1998). This statement, often referred to as the Lofthus syndrome, represents a major obstacle to media-sensitive construction projects. In such project environments, there are potential aggressive attackers, including journalists, politicians and other key players in the arena in question, taking note of what CEOs and other key personnel state that projects are expected to cost or when they are due to be completed.

Sooner or later, most development projects show discrepancies in cost or time compared with the initial budget or schedule. Major deviations are searched for or even created by attackers. Uncertainties, changes to project scope, etc. are ignored. Managers of media-sensitive projects therefore continuously have to watch out for attacks and be prepared to answer for any deviation, real or artificially designed by attackers.

The kind of project environment described here represents a major driver in the direction of closed projects and the use of A strategies. The ultimate choice is either to stay within the Juran dilemma and run your project safely and be protected, or go for an improvement in project value and fight the attackers from day one onwards.

If projects are to be run in a continuous-improvement mode, the environmental threshold, illustrated by the Juran dilemma, the Kaasen paradox and the Lofthus syndrome, represents a major obstacle that has to be taken seriously. New project management competencies are required, of which the proper understanding, managing and communicating of uncertainty as opportunities are major drivers for enhancing project value.

What is uncertain and why?

In ancient times uncertainty was closely related to games, mostly involving money. Nowadays money-games are played in financial markets, and only to a small extent at casinos. The development of a theory of uncertainty is therefore, not surprisingly, rooted in games or financial markets, or, expressed more generally, in decision-making situations concerning the deployment of money.

A comprehensive review of the historical development of uncertainty theory as applied to economics and finance is presented in Bernstein (1996).

In economics and finance the handling of uncertainty is highly quantitative. The variation and spread of key state variables are monitored continuously, and the data are analysed systematically by use of modern statistical methodologies. Likely consequences of the uncertain outcomes of future events are highlighted before decisions are made. Investments in high-uncertainty options are risky, i.e. the investor must be aware of the consequences of possible outcomes. The higher the uncertainty and the worse the potential negative consequences, the higher the risk of the investment.

The investor caters for risk by adding a risk premium to the expected value of the investment. The investor may, however, handle risky options differently from adding premiums. Spreading the risks by investing in a number of smaller options rather than in a single big one (diversification) is by far the most popular strategy.

In engineering disciplines, uncertainty is not directly tied to money as in economics and finance, but rather to variables of a technical nature. Uncertainty in environmental forces such as winds, waves and currents has a direct influence on the calculation of structural members. Structures, such as buildings and harbours, cannot be sized to withstand expected values of the forces acting on them; they must be designed to some standard, i.e. designed to withstand forces occurring more frequently than once in, say, 50 years.

The engineer and the economist typically encounter uncertainty issues with a number of highly quantitative techniques, and they often add premiums to safeguard themselves. A social scientist, however, has a somewhat different approach to uncertainty. While the economist focuses on variables in the money market and the engineer on technical variables, the social scientist's concern is the dynamics of the life world. The life world consists of people continuously communicating, acting or resting. The challenge of the social scientist is not so much to observe what is happening as to understand what is observed and why people behave as they do.

Engineers and economists make models that explain causes and effects; the social scientist makes models that interpret observed phenomena. The methodologies applied by the social scientist therefore are likely to be more qualitative than quantitative.

In projects engineers, economists and social scientists work side by side, their various concerns are highly interrelated, interwoven and integrated. Typically, uncertainty is of vital concern to them all in major decision situations. As a project, by definition, may be viewed as a social construction of reality, characterised by a strict goal orientation, uniqueness, low frequency and cross-functional virtual teams, project uncertainties are likely to be greater and comprehend a larger variety of issues compared with tasks of a high-frequency repetitive nature carried out within a vertically integrated organisational structure enforcing unity of command.

Uncertainty in situations typical of the management of projects

Over the project lifetime, decisions are made concerning go/no go choices, the development and implementation of execution strategies, control measures, and handover rituals for the project objective. At this point an important distinction has to be made between:

- *type 1 decisions*: the few but vital go/no go decisions made by the project sponsors
- *type 2 decisions*: the many, important managerial decisions related to planning and control, organising, and leadership (ongoing decisions made by project management)
- *type 3 decisions*: the myriad of decisions made daily by project team members, often related to the project workflow.

The type 1 decisions are acknowledged by a number of authors on project risk management, and appropriate models have been suggested to deal with these issues. The Project Risk Analysis and Management Guide (PRAM) is a good example of such a model.

The type 2 decisions have been known to project professionals for a long time. A large number of methodologies have been suggested over time. The state of the art today is a combination of a risk reduction procedure and a toolkit to calculate variation spreads in key variables related to the project workflow. However, it can be argued that the prevailing methodologies are inadequate for projects where the goal is to improve or radically change perceived best practices.

The type 3 decisions have mostly been ignored or have been considered trivial. Most project models treat uncertainty originating from decisions at the workflow level as a random variation in time, productivity or cost.

At this stage, it is convenient to broaden our understanding of the nature of uncertainty. Rather than try to select one of the many competing schools of thought regarding uncertainty management, it is more useful to consider a framework sufficiently broad to encompass all prevailing schools. A classification of uncertainty into four broad categories, represented by a 2 × 2 matrix would be appropriate (Figure 8.1). One axis expresses whether the state of uncertainty is considered closed or open, the other whether data describing the uncertainty state are known or unknown.

Deterministic uncertainty

This is used to express uncertainty in closed game situations where all information about possible outcomes, i.e. the number of possible outcomes, their values and their probability of occurrence, is fully known. This is the kind of situation we encounter in casinos. The

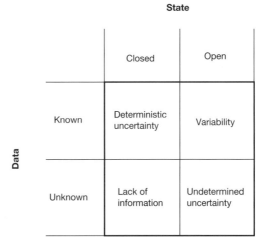

Figure 8.1. Classification of uncertainty

algorithm of the game is given, hence the notion of deterministic uncertainty.

In spite of the fact that real-life situations could hardly be said to mirror the spinning wheel of a casino, the 'Monte Carlo' or 'casino algorithm' is widely used to calculate uncertainty in real-world situations. The reason for this is twofold:

- Uncertainty in well-known repetitive situations may be considered as close to the spinning wheel of a casino.
- Project uncertainties are related to random variation in the state variables; any impact originating from corrective measures by the project manager is ignored.

The deterministic-uncertainty perspective is considered relevant to projects of a closed nature, or, more precisely, the prevailing project risk management models assume it is.

Lack of information

This is an acknowledgment of uncertainty rooted in missing pieces of information, which may result in an incomplete, ambiguous picture of the state of a project.

The prevailing project strategies encourage searching for more information until unambiguous interpretations of given facts can be obtained – presuming that the cost of acquiring new information adds value to the project.

A typical example is the case of an archaeologist. Carefully, bits and pieces of evidence are recovered and then an attempt is made to try to combine the given pieces in any way that make sense. Analogously, a civil engineer samples soil tests before designing foundations for a new building. The challenges are similar in the sense that uncertainty in critical parameters exists. The events that produced the ancient specimen and the type of soil at the building site have happened. Though it is known that something happened many years ago, the precise details of exactly what happened might be unknown.

In some instances, such as the challenge of the civil engineer, the number of tests and analyses may be increased until a sufficient degree of certainty is achieved. In the case of the archaeologist, however, some pieces of a specimen may be lost for ever, and hence a complete picture may never be constructed.

A first approach to the lack-of-information type of uncertainty may be illustrated by a puzzle. Assume a number of pieces are missing. What we do not know for sure is:

- how many pieces there are altogether (or how many samples, of what kind, are required)
- whether any of the known pieces belong to a different puzzle (are relevant information/data relevant to the phenomena at stake?)
- what pieces are missing (what should we do to complete the picture?).

A second approach to closed–unknown uncertainty may focus on uncertainty resulting from interpreting available data (ambiguity).

The lack-of-information way of thinking is frequently applied to non-repetitive projects exposed to high uncertainty of a closed nature – i.e. uncertainty that can be reduced at the expense of money and time.

Variability

This is the quantification of uncertainty in state variables using a combination of historical data and (inter-)subjective probabilities. The approach differs from the concept of the deterministic type of uncertainty as the number of outcomes, their values and their probabilities of occurrence are not given objectively any more, but rather are 'guessed' on the basis of professional judgement and intuition supported by a varying number of historical facts.

The variability perspective is primarily applied in planning and scheduling techniques. Typical examples of its application are estimations of hours worked, cost and activity duration. Data from similar past projects are recorded and treated statistically to produce estimating norms for future projects of the same kind.

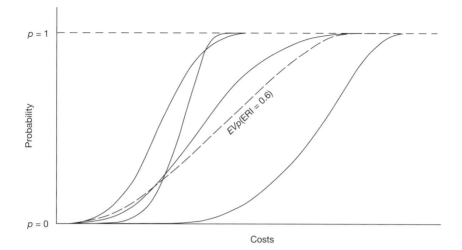

Figure 8.2. Use of inter-subjective probabilities

In planning and scheduling techniques, estimates of cost and time are represented by their expected values. Uncertainty is either ignored (resulting in deterministic models) or represented through a statistical distribution. In the latter case uncertainty, estimated at the activity level, has to be aggregated to the overall project level. This is done either through a full stochastic computation of the project activity network (Monte Carlo simulation) or through a short-cut technique such as PERT.

The traditional assumption made in applying these techniques is that the future will be a repeat of the past and that no learning takes place. Estimates of uncertainty, though, are frequently based on subjective assessments as required to complete historical observations. This approach may be improved by replacing subjective probabilities with probabilities of an inter-subjective nature (Figure 8.2).

Figure 8.2 illustrates the professional judgement of four experts. Their judgements differ. Our interest is partly in making an aggregated judgement and partly to assess by how much the various judgements differ. The aggregated judgement is represented by the dashed curve in the diagram, *EVp*. The curve was calculated by means of an ordinary regression analysis assuming that the experts' judgements were observed data. The dispersion – the regression coefficient – is referred to as the ERI (epistemic reliability index), an expression inspired by Gardenfors and Sahlin (1989). An ERI equal to 1 means that the judgements of the experts coincide, i.e. the experts are in an inter-subjective agreement An ERI equal to 0 indicates that the experts are in complete disagreement. Generally,

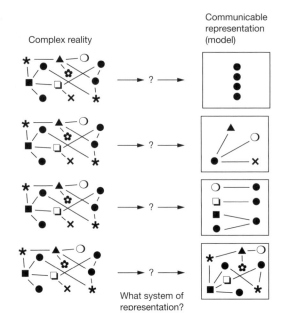

Figure 8.3. Modelling of undetermined uncertainty

ERI = 0 indicates that there is no justification for any assessment of a quantitative nature, and hence a qualitative assessment exploring a variety of optional judgements is encouraged.

A great number of variability models focus on uncertainty in state variables; decision variables are mostly ignored. Consequently, even advanced Monte Carlo simulation models are stupid or have an IQ equal to zero (Jordanger, 1990). No project is run without a manager of some kind interfering with nature as appropriate in order to keep the project on the planned track and to correct deviations as they occur.

Undetermined uncertainty

This requires the acknowledgment of open–unknown uncertainty. Undetermined states are complex by nature and require complex modelling. Reality may be modelled in a number of different ways, some of which are complex by design (Figure 8.3) (Genelot, 1992).

The nature of complexity

The term 'complexity' is often used to describe the fact that something is complicated (Lissac and Roos, 1999). In this chapter, the term 'complexity' is used to mean that something may happen that cannot be explained by sheer coincidence or an obvious cause–effect relation. Complexity means open interaction between actors (or, in general, interaction between components). Actors will group and regroup in a way that is neither coincidental nor given. Within the group, the actors may communicate and influence each other in ways that cannot be predetermined. The results of the group's action can therefore not be established with certainty before the action has been performed.

A more thorough description of this view of complexity is given in Eve *et al.* (1997). Below is an extract from this view of complexity (Eve *et al.*, 1997):

> Complexity is not a property just of the number of component parts or even the direction of their relationship, but of the variety of their interactions and thus the possibility to align into many different configurations.

> No matter how large the number of components, if there is no potential for components to interact, align, and organize into specific configurations of relationship, there is no complexity.

> Even if components are organized, but within such completely confining arrangements that no further possibilities for variety in interaction are left open, there is no complexity. Complexity, therefore, has to do with the interrelatedness of components as well as their freedom to interact, align, and organize into related configurations.

> The more components and the more ways in which components can possibly interact, align, and organize, the higher complexity.

From the above we learn that the category of undetermined uncertainty is complex by nature. The other categories, in the way they are defined, are not complex. Thus, not all situations are complex; however, all complex states are uncertain (undetermined uncertainty).

During the last few years, complexity has been an issue of great interest. In particular, research institutes in Los Alamos, Santa Fe and Brussels, embracing researchers such as Feigenbaum, Gell-Mann, Prigogine and Stengers, have made major contributions to the emerging science of complexity. Some new terms are 'chaos', 'bifurcation', 'fractals' and 'attractors'.

Later, scientists from other fields became interested in complexity. In particular, research performed by social scientists on the complexity of organisations ought to be of interest to project management professionals. This research often has dynamic system theory as its starting point. Through extensive simulations, a number of essential principles that embrace the so-called complex human systems have been identified (Stacey, 1996):

(1) Complex systems often produce unexpected and counter-intuitive results.

(2) In complex systems, where there is a non-linear relationship with positive and negative feedback, the links between cause and effect are distant in time and space. This means that it is extremely difficult to make specific predictions of what will happen in a specific place over a specific time period. Instead, quantitative simulations on computers can be used to identify general qualitative patterns of behaviour that will be similar to those we are likely to experience, although never the same.

(3) Complex systems are highly sensitive to some changes but remarkably insensitive to others. Complex systems contain some influential pressure points. If we can influence those points we can have a major impact on the behaviour of the system. The trouble is that these are difficult to identify. More usually, it seems, complex systems are insensitive to changes and indeed counteract and compensate; that is, to move to stability, it is necessary to change the system itself rather than simply apply externally generated remedies.

It should be noted that these principles emphasise that a complex social system, such as a project organisation, cannot be controlled from the outside.

Points (2) and, particularly, (3) draw our attention to whether the system gives negative feedback (the system seeks to restore stability) or positive feedback (the system is successively brought further and further away from stability). Positive and negative feedback usually act simultaneously, and it may be difficult to predict whether a system will settle down or move towards instability. Negative feedback may be desirable to ensure that the organisation works towards the realisation of common goals; positive feedback may be desirable to ensure creativity and innovation.

In Figure 8.4, a differentiation is made between an open and a closed structure. By 'closed' it is meant that the project organisation is formed by the management of the corporation. The group members have no dominant influence on how many or who will participate in the project.

By an 'open' structure it is meant that the project group forms itself. The size, structure and membership are determined along the way as the way the work gets started. The group establishes its authority and its internal and external lines of communication as appropriate.

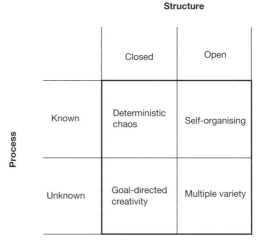

Figure 8.4. Differentiation between an open and a closed structure

Further, a differentiation is made between a known and an unknown process. By 'known' it is meant that we have a rather comprehensive understanding of work processes as well as managerial processes. By 'unknown' it is understood that we apply processes without being sure where they will lead us.

Deterministic chaos

The theory of chaos says that, when a deterministic non-linear system is driven away from the peaceful state of stable equilibrium towards the hectic state of explosive instability, it passes through a stage of bounded instability in which it displays highly complex behaviour, in effect the result of the system flipping randomly between positive and negative feedback (Stacey, 1996).

Deterministic chaos is likely to occur when a system is pushed towards a stage of instability. On its way, the system passes through a state of bounded instability in which it displays complex behaviour, i.e. the system never behaves in a regular way that leads to equilibrium. The behaviour is bounded in the sense that the system's successive states are to be found within a restricted basin – an attractor – of possible positions. What we find when we have deterministic chaos is that the system's successive states are not simply anywhere but rather are to be found within a restricted set within the range of possible positions (Byrne, 1998).

The prime focus of the deterministic-chaos perspective is the acknowledgement that small changes may drive systems far from equilibrium into chaotic behaviour. Sets of deterministic, time-differential equations in a model of change processes yield apparently indeterministic, unpredictable results. The mathematical equations used are non-linear, meaning that the dependent variables change at rates that are not simply first-order powers of changes in the independent variables. Furthermore, the corresponding coefficients are not dependent exclusively on changes of the values of the independent variables but also depend on changes in the boundary conditions or parameters of the model. When the model parameters are dynamic, these dynamics map from initial conditions to properties at later points in time. This is what is meant by extreme sensitivity to initial conditions (Eve *et al.*, 1997).

The deterministic-chaos perspective explains why extraordinary results achieved in one project may be impossible to reproduce.

Goal-directed creativity

Innovation and creativity occur where information from the chaotic external world meets the structured information of the internal world. Creativity is the process of making new meanings in the combination of these two domains (McMaster, 1995).

When procuring a service from someone in a contracting industry, you want innovation and creativity to be part of the package you buy. But people in a contracting industry will have a litany of reasons as to why they cannot spend time or money on innovation and creativity. The contracting business has been developed for production, not for learning or for innovation. To foster creativity, changes to the current structures and processes are required.

In order to understand what is meant by creativity in such a context, McMaster suggests the following conditions (McMaster, 1995):

> Flexibility emerges out of a structure. Specifically, it emerges out of a complex adaptive system that can be enhanced or inhibited by design. The elements of this structure are:
>
> - boundaries intentionally created that are meant to be temporary and changed regularly
> - accountabilities for structures that are outside those structures and therefore only have a pragmatic interest in their continuation
> - communication structures cross boundaries that come into existence with every creation of a boundary, as well as the ability to demand such communication outside the structure
> - measures of speed of change in relationship to registered or perceived changes in circumstances.

The suggested conditions tend to view an organisation as a living organism. Therefore studies of living systems may be useful for organisations. Theories from the world of evolutionary ecology offer interesting perspectives that allow us to see an organisation as a co-evolving system within a larger complex system and to see that the system, as a whole, is also an adaptive system.

In a complex adaptive system, such as an organisation, agents interact in a manner that constitutes learning (Gell-Mann, 1994). In organisations, the learning process is conscious and intentional.

Self-organisation

A network organisation is a market mechanism that allocates people and resources to problems and projects in a decentralised manner. As in the case of a market, efficiency is assumed. In a network organisation a novel problem is routed by the shortest path to the right people, while in a hierarchy a novel problem takes long paths by wending its way through channels established for familiar (routine) problems (Eccles and Crane, 1992).

A network organisation can flexibly construct a unique set of internal and external linkages for each unique project (Baker, 1992). The open communication patterns make network structures self-organising. The important distinction between self-organising teams and self-managed teams has been noted (Stacey, 1996, p. 333):

> Self-organising is a fluid network process in which informal, temporary teams form spontaneously around issues, while self-managing teams are permanent or temporary but always formally established parts of a reporting structure.

> While top managers cannot control self-organising processes – they can only intervene to influence boundary conditions around them – they can install a structure of self-managing teams and control them through the rules that govern how they are to operate.

Self-organising perspectives are considered useful in open projects, where it is important that stakeholders form loosely coupled structures in order to align their various project objectives.

Multiple variety

This reflects the multitude of composite views that one may take in the modelling of projects, i.e. the freedom of choice concerning stakeholders' interests, organisational structures (formal and informal, intra- and inter-organisational, judicious, and virtual), and working and managerial processes and contexts.

The multiple-variety perspective is fundamental to contemporary project practices as it underpins the importance of project constructs in two ways:

- a stakeholder may choose between numerous constructs, of which more than one are considered feasible
- different stakeholders are likely to come up with different project constructs.

Managing uncertainty

Given the answer 'use the RAMP guide', what is the question? Does one size (guide) fit all (projects)? Is the RAMP guide applicable to all projects, or at least to most projects? If the answer is yes, is the recommended approach in the guide the only solution, or if not, is it the best one?

Suppose the RAMP guide recommendation is compared with the investor's choice of putting money into a savings account. In both cases risk reduction and increased predictability are the rationale for the selected option. But in what kind of situations are these actions the preferred strategy?

As shown in Figure 8.5, Epci expanded the Obeng space to four ideal types of project ambitions (Hetland, 2002):

- *status quo*
- improvement
- radical change
- *quo vadis* (searching for direction).

There are a number of related project strategies by which values are being enhanced. Four main families of strategies (Figure 8.6) will be addressed as an illustration of the variety of approaches

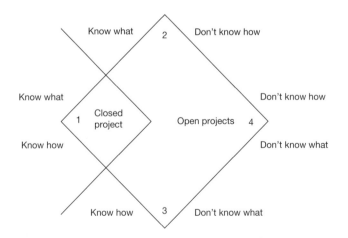

Figure 8.5. The four ideal types of project ambitions

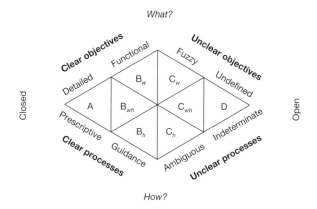

Figure 8.6. The four types of strategy

available to contracting parties. The strategies are vastly different as facilitators for change and learning. For the purpose of reference, the selected strategies are referred to as:

- A: clear, detailed objectives and clear, prescriptive processes
- B: functional objectives and guiding processes
- C: fuzzy objectives and ambiguous processes
- D: undefined objectives and indeterminate processes.

The choice of strategy is dependent not only on the particular kind of project challenge, but also on the intended course of action. Type A strategies used to prevail, and were applied to all kinds of projects. Type A strategies are particularly appropriate to cope with closed projects, and focus on copying best practices and extensive control such that actual processes run according to detailed specifications and predefined procedures. Projects of this type dominate the recommendations in the RAMP and similar guides.

Recently, however, many organisations have been exposed to B strategies and use them for improving products and processes. In oil-related projects in the North Sea, some experiments have turned out very successfully; others have had disappointing results and considerable cost overruns. The cluster of the emerging best-practising organisations are climbing the B strategy learning curve(s); the champions of yesterday are stuck on the top of the A curve, unable to shift to alternative paradigms.

High-tech organisations are experimenting with C strategies; they have to. Complex strategies (B and C) applied to complex projects have far higher value potentials than do linear strategies (A) applied to simple projects. No-one consciously *plans* to apply D strategies to

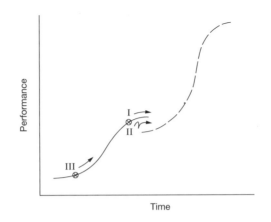

Figure 8.7. Status-quo-*type projects*

projects, but some may find themselves in that position to their surprise. Projects may be so badly managed, or hit by severe external disturbances, that D strategies are the only way out of a situation that may be described as 'don't know where to go, but can't stay here' (Hastings *et al.*, 1994).

Status-quo-type projects

A great number of projects are trivial in the sense that we know what to do and how. It is just one more project, similar to the ones that have been carried out for decades (Fig. 8.7). Procurement practices have been developed and refined over the years. Learning has been incremental. At the outset, all requirements for skills and knowledge are known. Working skills have been accumulated and are transferred to new generations of employees. Typically, work processes are prescriptive, allowing a carefully designed step-by-step sequencing of activities. The focus is on copying or exploiting best practices, avoiding errors such as rework, excess waiting time and similar phenomena that affect time and productivity.

Value-enhancing procurement strategies applicable to *status-quo-*type projects are typically linear, contract structures are hierarchical and work processes are sequential. The overall philosophy is to reduce uncertainty to a minimum before major contracts are let. Project work is split into separate engineering and construction contracts. No construction contract is let unless a comprehensive definition of the work scope has been produced. A no-change policy (and hence a non-learning) policy is enforced.

The foci of the value-enhancing mechanisms in A strategies are:

- avoid errors
- discourage changes
- encourage efficient work processes.

This strategy ignores uncertainty. Uncertainty either is a non-issue or is eliminated through the preparation of complete definitions of the project work scope, schedules and resources. The problem is that *all* projects are thought to have the same characteristics, with the same corresponding processes, and therefore deserve the same tools and techniques.

These processes and their respective definitions are:

- *Initiating processes*: authorising the project or phase.
- *Planning processes*: defining and refining objectives and selecting the best of the alternative courses of action to attain the objectives that the project was undertaken to address.
- *Executing processes*: coordinating people and other resources to carry out the plan.
- *Controlling processes*: ensuring that project objectives are met by monitoring and measuring progress regularly to identify variances from the plan so that corrective action can be taken.
- *Closing processes*: formalising acceptance of the project or phase and bringing it to an orderly end.

Improvement-type projects

For B strategies to succeed, learning is essential. B strategies do not aim at reducing uncertainty; uncertainty is, rather, deployed as a means to enhance value. Project objectives are not completely determined; they are often expressed in terms of functional requirements. Furthermore, work processes are not prescriptive; more open guidance documents are used. In sum, the B strategies encourage suppliers to select their own objectives and procedures within the broad statements provided by the principal or client. Suppliers may thus choose their own way of working, capitalise on their own capabilities and develop their own core capabilities as complementary to those of their clients.

To provide an example, this strategy may be implemented when the ambition of the client (and hopefully the common objective) is to improve project cost by 40% by application of the Norsok principles (a semi-closed, semi-open project challenge). Relatedly, the supplier has to command competencies in managing B_{wh} projects. This implies that the project requires knowledge and experience presently unknown to the organisation. Thus the organisation will have to either initiate internal learning processes or acquire sufficient and complementary knowledge from partners. In other words, intra- or inter-organisational learning and/or experience transfer is a critical success factor.

On the basis of a differentiated need for knowledge access, two major schools in the application of B strategies may be delineated: the EPC (engineer, procure and construct) and the PAHA (project *ad hoc* alliance) contracting philosophies. In the EPC school, suppliers are sheltered from 'client disturbances'. Major achievements in terms of project effectiveness and efficiencies are expected as contractors provide engineering services for their own construction work. To succeed, the company needs to excel in integrating knowledge from different corners of the organisation. Integration efforts within the firm are seldom straightforward. Internal hegemonies and values are challenged openly. This having been said, cross-functional projects provide the most likely organisational design for success.

In this context, projects offer ample opportunities to develop a common language (formally as well as informally), joint problem solving, and a common understanding and appreciation of goals, rules and procedures. All these mechanisms serve as instruments for coordinating and integrating knowledge. In general, firms are regarded as having certain advantages as institutions for supporting knowledge-integrating mechanisms. However, similar integration may be achieved through coordination through inter-organisational collaboration.

In the PAHA school, the potential for improving project effectiveness is even larger, as the client may source complementary knowledge and technologies from suppliers, and thus leverage the value-enhancing processes of projects. The success of the PAHA concept, though, is sensitive to the degree of mutual trust and respect between the alliancing partners. These processes lend themselves to studies of projects over time, to see how relations and trust may evolve during the lifetime of projects, or through repeated projects where the same partners are involved.

Experience may, in this context, be a dangerous teacher. Prior experience may not be relevant at all, a fact that may be hard to acknowledge for experienced project members. Individual cognitive limitations and organisational politics are major barriers to improvement projects. Experience may produce the wrong solutions and represent serious constraints on the project's ability to learn. The relevance of experience will of course depend on the degree of improvement required. This leads us to the next generic project category.

Radical-change-type projects

In a tumultuous world, a growing number of projects are seen in the radical-change area. In the North Sea context, typical examples are goals to cut cost and time by a half without affecting the quality of project work, or to apply a brand new technology. The first is referred to as a C_w challenge, the latter as a C_h. Similarly to the B strategies, the two challenges may be merged to give a C_{wh} strategy. In these strategies, the exploration of new ideas will be in high demand. Avoiding competence traps, and consciousness of path dependencies are major issues.

A typical feature of C strategies is that the project challenge is unlikely to be solved inside the given area of freedom to act. The boundaries of the contract or the standard mode of project organising has to be challenged. Creative thinking replaces logical thinking. Linear work patterns such as hierarchical organisation structures and sequential working procedures are replaced by complex ones.

In order to cope with low predictability and high complexity, radical-change strategies apply a 'go as you please' pattern. The first step is to enforce creative thinking by various ways of blocking traditional thinking. Examples of such procedures are to be found in the concepts known as lateral thinking and synectics.

Another frequent feature of the first-step processes is to let two or more groups work in parallel on the same problem. Instead of one supplier you assign several. Examples of such efforts are seen in architectural competitions, concept competitions, etc. Suppliers may be fully paid for services rendered, but more frequently they are only paid for work undertaken. The client will, however, remunerate the winning concept. The successful supplier may, further, be selected as the main supplier for the next steps of the project. For the next step in the project process, a C strategy may be applied again, or a B one may be considered. Finally, for the last step, even an A strategy may be applied.

Radical-change type projects are often broken down into a number of sub-projects (project clusters of the chain or programme type). Each sub-project will then be assigned to the appropriate location in the VEPS project map. Through this approach, complexity is reduced and C strategies are applied to the radical-change positions only.

The starting point for the goal and context challenge (zone C) is an *a priori* goal and context. Rather than simply accepting these as given requirements, this position advocates that the goal and the context are continuously challenged to see if a change may enhance the project value. The procedure is dynamic and requires an open and demanding dialogue between key project stakeholders.

Radical-change projects may benefit from a knowledge infrastructure that permits these two-dimensional processes to flourish. Accessing and disseminating tacit knowledge is realised by sharing experiences and by reflection on action. As reflection on action usually requires shared time and space, these infrastructures tend to be more expensive, time-consuming and difficult or costly to replicate and transfer.

Quo vadis **projects**

Though the potential for value enhancement through the design of situations leading to chaos some day may be considered appropriate, I know of no project deliberately launched as a *quo vadis* project. More likely, such a project situation occurs as a result of actors and organisations losing control of a project for whatever reason. Hence, *quo vadis* situations often need a quick response. Consequently, the fact that a client is under severe strain characterises the procurement practices. Services have to be provided immediately, there is no time for linear contracting procedures. The overriding supplier selection criteria are:

* Who may help us out of this situation?
* Who is available right now? And, possibly,
* Which supplier is known to possess a flexible organisation?

Here, creative groups are working in parallel under tremendous time pressure. They do not know the solution, and hardly know where to start looking for one. Price is often a non-issue; how can you bid for work in a situation in which you know neither what to do nor how? And the client has no time to waste on negotiating the contract.

At this stage, it is time to bring all of the above together. As stated above, four ideal types of ambitions have been presented:

* *status quo*
* improvement
* radical change
* *quo vadis*.

The first is trivial in the sense of uncertainty management. In the second and third cases, faced with an improvement or radical-change ambition, we start by asking what we want to improve or change. By answering that question, we indirectly answer the question of what we do not want to improve or change but instead keep similar to some known or standard platform. The non-improvement

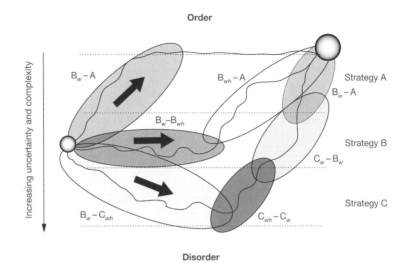

Figure 8.8. The value of uncertainty and complexity

and no-change elements of the project are often referred to as the project platform. To execute the platform elements, we simply apply the known best practice.

The elements to be improved or changed must be challenged in particular ways in order to make the improvement or change come through. Techniques derived from total quality management and from value engineering and management may help to achieve improvement. Techniques deriving from lateral thinking, synectics or similar creative processes may help to achieve change. Prior to deciding on a specific route to improvement or change, however, we should consider the environments relevant to the project. The following ideal types of environments may be relevant:

- (− −) aggressive attack
- (−) critical
- (0) neutral
- (+) supportive
- (+ +) proactive, demanding dialogue.

If the project is aggressively attacked or exposed to critical assessment from relevant environments, a linear–reductionistic strategy, previously referred to as a BA strategy (or CBA strategy for radical-change elements) may be the only viable approach. Other routes, illustrated in Figure 8.8, may expose the project to unacceptable risk levels, including regret risks.

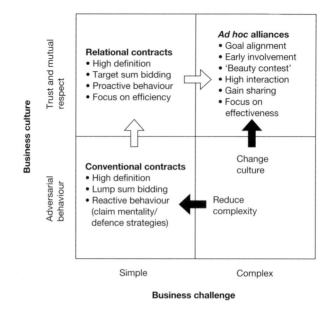

Figure 8.9. Impact of project environment and uncertainty (Hetland, 1996)

Proactive project environments encourage project teams to use uncertainty and undetermined states as freedom to act to facilitate a project process that aims at continuously enhancing the project value. The impact of project environment and uncertainty on the selection of an appropriate value-enhancing project strategy is illustrated in Figure 8.9.

Concluding remarks

Obviously, there must be a reason behind the reductionistic/deterministic approach to project and risk management. The rationale may be twofold;

(1) Unpredictability of project states is not acceptable, or at least is highly unwanted.
(2) The project stakeholders are simply not aware of the fact that alternative strategies, in certain situations, may produce higher project value.

Indirectly, the question to the original answer 'use the RAMP guide' has now been answered. The guide is very appropriate to handle situations where it is essential to have tight control in order to avoid scope increases, time delays and budget overruns.

There are a number of situations where project success is equal to being in control of scope, time and cost discrepancies. However, there are situations that are different, some of them vastly different – where success is related to expected value added beyond what can possibly be produced by means of reductionistic/deterministic approaches. Projects may not only exploit the available information, but also explore the potential of new knowledge and novel solutions. It is in this kind of situation that the management of uncertainty is of vital importance.

Bibliography

Baker, W. The network organization in theory and practice. In: Nohria, N. and Eccles, N. R. G. (eds), *Networks and Organizations*, chapter 15. Harvard Business School Press, Boston, 1992.

Berstein, P. L. *Against the Gods. The Remarkable Story of Risk*. Wiley, New York, 1996.

Byrne, D. *Complexity Theory and Social Sciences*. Routledge, London, 1998.

Chapman, C. *Proceedings of the Epci Annual Conference*. Hamburg, 1998.

Chapman, C. and Ward, S. *Managing Project Risk and Uncertainty*. Wiley, Chichester, 2002.

CRINE. *Cost Reduction Initiative for the New Era*. UKOOA, London, 1994.

Eccles, R. G. and Crane, D. *Doing Deals: Investment Banks at Work*. Harvard Business School Press, Boston, 1988.

European Institute of Advanced Project and Contract Management. *Proceedings of the Inaugural Conference*. Stavanger, 1994.

European Institute of Advanced Project and Contract Management. *Proceedings of the Annual Conference* Hamburg, 1998.

European Institute of Advanced Project and Contract Management. *Proceedings of the Annual Conference*. London, 1999.

European Institute of Advanced Project and Contract Management. *Value Enhancing Project Strategies*. Epci, Stavanger, 2000.

Eve, R. A. *et al*. *Chaos, Complexity and Sociology*. London, Sage, 1997.

Fevang, H. J. and Hetland, P. W. Exploring the value of project complexity – beyond lump sum contracting. In: Kähkönen, K. and Artto, K. A. (eds), *Managing Risks in Projects*. E & FN Spon, Helsinki, 1996.

Gardenfors, P. and Sahlin, N. E. (eds). *Decision, Probability and Utility*. Cambridge University Press, 1988.

Geddes, M., Hastings, C. and Briner, W. *Project Leadership*. Gower, London, 1993.

Gell-Mann, M. The Quark and the Jaguar. Freeman, New York, 1994.

Genelot, D. *Manager dans la Complexite*. INSEP, Paris, 1992.

Hastings et al. The role of projects in the Strategi Prosess. In: Cleland, D. I. and Gareis, R. (eds), *Global Project Management Handbook*. McGraw-Hill, New York, 1994.

Hetland, P. W. Målsøkende strategier og kontraktsformer for gjennomføring av komplekse utbyggingsprosjekter. PhD thesis, AUC, Aalborg, 1995.

Hetland, P. W. Best Contracting Practices. Projektmanagement Tag Österreich, Wirthshaftsuniversitet Wien. Presentation. Vienna, 1996.

Hetland, P. W. Value enhancing procurement strategies. *Proceedings of the European Construction Institute Annual Conference on Partnering in Europe.* Milan, 2000.

Hetland, P. W. Major trends and paradoxes in managing construction projects. In: Brochner, J., Josephson P.-E. and Larsoson, B. (eds), *Construction Economics and Organization.* Chalmers, Gothenburg, 2001a.

Hetland, P. W. Fremveksten av en situasjonsbestemt prosjektledelse-tradisjon. In: Gottschalk, P. and Welle-Strand, A. (eds), *Læring gjennom økonomi, system og prosjekt.* NKI forlaget, Oslo, 2001b.

Hetland, P. W. Lecture, European Programme for Project Executives. Epci, Stavanger, 2001c.

Hetland, P. W. A contingency approach to managing complex projects. *Proceedings of the PMI European Conference.* Cannes, 2002.

Jordanger, I. 1990. A humanised aproach to intelligent risk management. In: Proceedings, vol. 2. Management by projects. 10th Internet World Congress on Project Management. Vienna, 1990.

Lissack, M. and Roos, J. The Next Common Sense. Brealey, London, 1999.

Lofthus. Lecture, European Programme for Project Executives (EPPE). UMIST, Manchester, 1997.

McMaster, M. D. *The Intelligence Advantage. Organizing for Complexity.* Knowledge Based Development, Isle of Man, 1995.

Nohria, N. and Eccles, R. G. (eds). *Networks and Organizations.* Harvard Business School Press, 1992.

Norsok. *Norsok, Hovedrapport og delrapport 3, Samarbeid mellom operatør og leverandør.* Utbyggings- og driftsforum for petroleumssektoren, Oslo, 1996.

Obeng, E. *Putting Strategy to Work.* Capstone, Oxford, 1996.

Obeng, E. The Obeng space for understanding projects and transferring experience between industries. *Proceedings of the Epci Annual Conference.* London, 1999.

Olje- og energidepartementet. *Kostnadsanalysen – norsk kontinentalsokkel.* Universitetsforlaget, Oslo, 1980.

Olje- og energidepartementet. *Analyse av investeringsutviklingen i utbyggingsprosjekter på kontinentalsokkelen, Utredning fra Investeringsutvalget.* OED, Oslo, 1999, Kaasen Report.

Østensen, A. *Proceedings of the Epci Annual Conference.* London, 1999.

Project Management Institute. *A Guide to the Project Management Body of Knowledge (PMBoK Guide).* PMI, Newtown Square, 2000.

Stacey, R. D. *Strategic Management and Organisational Dynamics.* Pitman, London, 1996.

UKOOA (United Kingdom Offshore Operators Association). The CRINE (Cost Reduction Initiative for the New Era) Report. Institute of Petroleum, London, 1994.

CHAPTER NINE

Management and corporate risk

T. Merna

Introduction

The increasing pace of change, customer demands and market globalisation all put risk management high on the agenda for forward-thinking companies. It is necessary today to have a comprehensive risk management strategy. In addition, the Cadbury Committee's Report on Corporate Governance (Cadbury, 1992) states that having a process in place to identify major business risks as one of the key procedures of an effective control system is paramount. This has since been extended in the *Guide for Directors on the Combined Code*, (Institute of Chartered Accountants, 1999). This report is referred to as the 'Turnbull Report' for the purposes of this chapter.

The Turnbull Report is a timely reminder of this, and is also an opportunity to review what an organisation has in place and to make the appropriate changes. Risk management can be considered as the sustainability of a business in its environment. In the past, large corporate failures have occurred where risk assessment was never even considered. Reichmann (1999) states, 'One of the most important lessons I have ever learnt, and I didn't learn it early enough, is that risk management is probably the most important part of business leadership'. However, organisations do need to be pragmatic. Taking responsibility for risk is often needed in order to gain reward. This is clearly stated in the Turnbull Report, which states that 'risk management is about mitigating, not eliminating risk'. Under the Turnbull Report and the Companies Act, the board of directors of a strategic business unit (SBU) has overall responsibility for and ownership of risks.

The Turnbull Report is not just about avoidance of risk. It is about effective risk management: determining the appropriate level of risk, being conscious of the risks you are taking and then deciding how you need to manage them. Risk is both positive and negative in nature. Effective risk management is as much about looking to make sure that you are not missing opportunities as it is about ensuring you are not taking inappropriate risks. Some companies will seek to

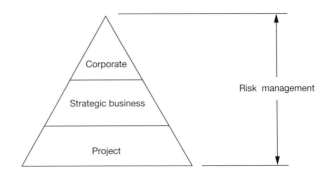

Figure 9.1. Levels within a corporate organisation

be more risk-averse than others. However, all should be seeking to achieve a balance between encouraging entrepreneurialism within their business and managing risks effectively. The Turnbull Report cites several principles of good corporate governance. Firstly, there are the directors. Factors controlled by directors include the board, the chairman, the CEO, board balance, supply of information, appointments to the board and re-election.

The purpose of the Turnbull Report is to guide British businesses and help them focus on risk management. Key aspects of the report include the importance of internal control and risk management, the maintenance of a sound system of internal control with the effectiveness being reviewed constantly, the board's view and statement on internal control, due diligence, and the internal audit.

The management of risk is one of the most important issues facing organisations today. This chapter will focus on risk management at the corporate, strategic business, and project levels within an organisation.

Figure 9.1 shows a typical organisation structure, which allows risk management to be focused at different levels. By classifying and categorising risk within these levels, it is possible to drill down and roll up to any level of the organisational structure. This should establish which risks the project is most sensitive to, so that appropriate risk response strategies may be implemented to benefit all stakeholders.

Figure 9.1 illustrates the corporate, strategic business, and project levels, which form the basis of this chapter. Risk management is seen to be integral to each level, although the flow of information from level to level is not necessarily on a top-down or bottom-up basis. In the authors opinion, risks identified at each level are dependent on the information available at the time of the investment, and each risk may be covered in more detail as more information becomes available.

Risk management

Risk management cannot simply be introduced to an organisation overnight. The Turnbull Report lists the following series of events that need to take place to embed risk management into the culture of an organisation:

- *Risk identification.* Identifying on a regular basis the risks that face an organisation. This may be done through workshops, interviews or questionnaires. The 'how' is not important, but actually carrying out this stage is critical.
- *Risk assessment/measurement.* Once risks have been identified it is important to gain an understanding of their size. This is often done on a semi-quantitative basis. Again, the 'how' is not important, but organisations should be measuring the likelihood of occurrence and the impact both in terms of image and reputation and in terms of financial impact.
- *Understanding how the risks are currently being managed.* It is important to profile how the risks are currently being managed and to determine whether or not this meets an organisation's risk management strategy.
- *Reporting the risks.* Setting up reporting protocols and ensuring that people adhere to such protocols is critical in the process.
- *Monitoring the risks.* Risks should be monitored to ensure that the critical ones are managed in the most effective way and the less critical risks do not become critical.
- *Maintaining the risk profile.* The need to maintain an up-to-date profile in an organisation to ensure that decisions are made on the basis of complete information.

Often, risk management forms part of an integrated management system along with quality management, planning, health and safety management and change management. In a competitive economy, profits are the result of successful risk taking. Against this background, the Turnbull Report endorsed by the London Stock Exchange in the same year, strives not to be a burden to the corporate sector, but rather to reflect good business practice. Companies should implement any necessary changes in a way that reflects the needs of their business and takes account of their market. As and when companies make those changes, they should discover that they are improving their risk management and, consequently, get a benefit that justifies any cost.

Corporate management

For the purpose of this chapter, corporate management is defined as *the management of the activities carried out by the corporate body and those organisations forming part of the corporation which utilise tools and techniques to aid decision-making processes.* Corporate management, often known as corporate strategy, is concerned with ensuring corporate survival and increasing its value not just in financial terms but also by variables such as market share, reputation and brand perception. Thus the scope of corporate risk management is wide-ranging, to support the corporate strategy. Corporate strategy is normally set through the implementation of a five-year plan.

A senior corporate manager owns the process and has the staff to resource the analysis and administrative activities. A board member champions the process, ensuring access to information and resources. A core group of corporate board members and SBU executives can draw additional input from stakeholders such as:

- shareholder representatives
- representatives from major customers, partners and suppliers
- external experts.

The scope covers the current markets and project portfolios of the SBU and also looks for potential new markets. Results fed back from the SBU are assessed along with changes and trends in international markets (customers, suppliers and competitors), legislation, politics and social attitudes.

The information used often comes from a range of sources, sometimes more than one; these may include:

- internally generated information
- corporate strategy plan
- corporate financial reports
- business unit financial reports
- feedback from business unit risk monitoring
- information from the public domain
- competitor, customer, supplier and partner financial reports
- benchmarking and forecasts from professional bodies, such as the Confederation of British Industry (CBI)
- research papers
- information from pressure groups
- government-generated information
- economic statistics and forecasts
- demographic and socio-economic trends
- White and Green Papers (UK Government)

- consultation on proposed legislation
- information purchased from specialist organisations, such as independent research analysts
- consumer trends
- technology forecasts
- information from past and present projects.

Corporate functions

Every firm needs a corporate mission. This mission encompasses the basic points of departure that send the organisation in a particular direction. McCoy (1985) states that the purpose of an organisation is the most important point of departure of strategy making, but also influential are the values embodied in an organisation's culture. Falsey (1989) believes values shared by organisation members will shape what are seen as ethical behaviour and moral responsibilities, and therefore will have an impact on strategic choices.

Other drivers influencing the corporation include where the corporation wishes to focus its efforts, and its competitive ambitions or intentions are an important part of the mission.

The corporate mission can be articulated by means of a mission statement; however, in practice not everything that is called a mission statement meets the above criteria. However, the author believes companies can have a mission, even if it has not been recorded on paper, although this will increase the risk of divergent interpretations throughout the corporate level.

In general, the corporate-level mission provides three important roles for an organisation. These roles are:

- *Direction.* The corporate mission should point the organisation in a certain direction. This is done by defining boundaries, within which strategic choices and actions must take place. However, by specifying the fundamental principles on which strategy must be based, the corporate mission limits the scope of strategic options, therefore setting the organisation on a specific course.
- *Legitimisation.* The corporate mission can convey to all stakeholders, on each level and outside the company, what the organisation is pursuing, and that these goals and objectives will add value to the company. By specifying the business philosophy that will guide the company, it is hoped stakeholders will accept, support and trust the corporate heads within the organisation, thereby generating support throughout the corporate, strategic business, and project levels.

- *Motivation.* In some cases, the corporate mission can go one step further than legitimisation, by actually inspiring individuals and different levels of the organisation to work together in a particular way. By specifying the fundamental principles driving an organisation, it can cause a 'corporate spirit' to evolve, generating a powerful capacity to motivate people over a prolonged period of time.

Within corporations, a concept that is often confused with mission is vision. A corporate vision is a picture of how the corporation wants things in the future to be. While a corporate mission outlines the basic point of departure, a corporate vision outlines the desired future at which the company hopes to arrive. However, the above corporate themes are very important considerations, and a great deal of time and effort must go into generating these at corporate level (David, 1989).

Corporate governance

At corporate level, an area that requires attention is that of who determines the corporate mission and regulates the corporate activities. This is corporate governance: who deals with the issue of governing the strategic choices and actions of top management (Keasey *et al.*, 1997)?

Corporate governance is concerned with building in checks and balances to ensure that top management pursue strategies that are in accordance with the corporate mission. Corporate governance encompasses all tasks and activities that are intended to supervise and steer the behaviour of top management. This is known as the corporate governance framework. This determines whom the organisation is there to serve and how the purposes and priorities of the organisation should be decided. It is concerned with both the functioning of the organisation and the distribution of power among different stakeholders. This is strongly culturally bound, resulting in different traditions and frameworks in different countries (Yoshimori, 1995).

The Turnbull Report discusses good corporate governance, including directors, as mentioned earlier. Factors controlled by directors include the board, the chairman, the CEO, board balance, supply of information, board appointments and re-election.

Every company listed on the London Stock Exchange should be headed by an effective board, which should lead and control the company. There are two key aspects of the top of every public company, these being the running of the board, and the executive responsibility for running the company's business. There should be a clear division of responsibilities at the head of the company which will ensure a balance of power and authority, such that no one individual has unfettered powers of decision.

The board should include a balance of executive and non-executive directors (including independent non-executives) such that no individual or small group of individuals can dominate the board's decision-taking. It should also be noted that there should be a formal and transparent procedure for the appointment of new directors to the board.

Tricked (1994) cites the common definition of corporate governance as 'addressing the issues facing board of directors'. Attention must, therefore, be paid to the roles and responsibilities of the stakeholders involved at corporate level.

There are three important functions at corporate level:

- *Forming function.* The first function is to influence the forming of the corporate mission. The task here is to shape, articulate and communicate the fundamental principles that will drive the organisational activities. Determining the purpose of the organisation and setting priorities among claimants are part of the forming function. Yoshimori (1995) suggests the board of directors can conduct this task by questioning the basis of strategic choices, influencing the business philosophy, and explicitly weighing the advantages and disadvantages of the firm's strategies for various constituents.
- *Performance function.* This function contributes to the strategy process with the intention of improving the future performance of the corporation. The task here at corporate level is to judge strategy initiatives brought forward by top management and to actively participate in strategy development. Zahra (1989) believes the board of directors can conduct this task by engaging in strategy discussions, acting as a sounding board for top management, and networking to secure the support of vital stakeholders.
- *Conformance function.* This function is necessary to ensure corporate conformance to the stated mission and strategy. The task of corporate governance is to monitor whether the organisation is undertaking activities as promised and whether performance is satisfactory. Where management is found lacking, it is a function of corporate governance to press for changes. Spencer (1983) believes the board of directors can conduct this task by auditing the activities of the corporation, questioning and supervising top management, determining remuneration and incentive packages, and even appointing new managers.

Hussey (1991) categorised the objectives/functions of a company as primary objectives, secondary objectives and the corporate goals that a firm wishes to achieve.

Primary objective

Profit is the prime motivation for all companies, and many managers argue that achieving profit maximisation is their prime function. However, in some cases the above may be untrue because no company is willing to do absolutely anything for profit. For example, few companies would be willing to work their employees into a state of physical and mental exhaustion. When dealing with customers, most purchases or transactions are likely to be repeated in the future; therefore looking for a high one-off profit will have an adverse effect on long-term profit.

Secondary objective

At corporate level, the secondary objective is a description of the nature of the company's business. At corporate level, the question should be asked, 'What is my business?' This can be answered at corporate appraisal; however, this is not an objective. To overcome this, the question 'What should my business be?' can be asked. From this information, the CEO and his/her immediate managers, such as the marketing, production and finance managers, can decipher 'where', 'when' and 'why' the company chooses a particular direction.

However, it must be recognised that every CEO has in his/her own mind 'where', 'what' and 'how' he/she wants the company to operate, regardless of company strategy.

Corporate goals

Goals are quantifiable objectives that provide a unit of measurement, from which the CEO can confirm that his/her strategies have been carried out. They are, therefore, more difficult to formulate than profit goals because profit goals are directly related to the strategies put in place. Goals are the landmarks and milestones which mark the selected path the company takes to reach the reference point (Handy, 1999).

Corporate landmarks and milestones should be quantifiable, allowing targets for each of the important company operations to be compared and, in the long run, achieved. There should be as many goals as it is practical to develop. There is little point in developing figures or targets that the company has no intention of looking at or that are of no relevance to the task.

The author cites a number of practical goals to be carried out as a governing 'meter' at corporate level:

- employment figures
- ratios describing shares of defined market (percentage)
- accounting figures such as liquidity ratio, or gearing
- minimum figures for customer figures
- maximum figures for hours lost in industrial disputes
- return on capital employed
- absolute sales targets
- a value for operational profit improvement
- staff turnover rate (lower targets each year, i.e. continuously improve employees' situation by listening).

Recognising risks

For real-world companies in viciously competitively environments, it is not good enough simply to protect the physical and financial assets of the corporation through a combination of good house-keeping and shrewd insurance and derivative buying. The pressure on margins is too intense and the vulnerability to volatility simply too great for that to be an adequate strategy for most companies, even small ones. The focus must shift to the far greater and far less tangible world of expectations and reputation, and thereby to sustaining investor value. Hence the inexorable rise of risk management and its sudden popularity in the boardroom (Monbiot, 2000).

Equity and credit analysts are increasingly focusing on risk and the quality of risk management within the companies they analyse, which is further sharpening focus in the boardroom. Analysts want to be able to tell current and potential investors that the corporate management knows what it is doing and that it is using its capital in the most effective manner possible, and that it is in control of its SBUs and consequently future profits.

Senior managements are increasingly using company reports and press departments to boast about their latest risk management initiatives and policies, but learning the vocabulary associated with risk management and simply slipping the words into glossy brochures does not constitute risk management. Corporations that want to report the stable, secure, socially responsible and ever-increasing earnings that investors and other stakeholders demand must take risk management seriously and put such words into practice (Parkinson, 1993).

In the corporate sector, more enlightened senior management have hired overall risk managers, more often than not promoted from the insurance management function. Here these individuals' core responsibility has normally been the identification, measurement and mitigation of risk, as well as arranging its funding when feasible and desirable. In many cases, these individuals have attempted to coordinate the risk management activities of other departments and to promote a risk management culture throughout the organisation.

Recent surveys of CEOs and risk managers in the UK, Europe and the US have shown constantly that the main perceived issues today are corporate governance; extortion, product tampering and terrorism; environmental liability; political risk; regulatory and legal risk; fraud; and a whole host of risks ushered in by modern technologies (Monbiot, 2000).

The causes of this shift in emphasis are of course many, varied and inextricably interrelated. But essentially, corporate and financial risk has grown in scale and complexity in tandem with the globalisation of the world economy. The globalisation of trade and the removal of barriers at national and international levels have led to a massive process of consolidation in all sectors as essentially uneconomic organisations, which previously relied on a combination of customer ignorance, lack of external competition and government assistance, have been forced to adapt or die.

In this global, relatively and increasingly service-dominated economic environment, corporate success increasingly comes to rely on two key drivers – perception and knowledge. Risk management is an integral part of these and a thorough understanding of the concept will drive an organisation one step further towards success. Companies must have the ability to source raw materials at a good price and turn them into a marketable product at a price that delivers a healthy margin. However, contingencies must be put in place, through the use of a complete, structured and up-to-date risk management system.

One major risk to corporations is from hostile bids. Corporations often increase their financial gearing to employ more debt than equity and thus make themselves less attractive to opportunistic takeovers. Shareholders, however, do not necessarily want too much debt, as debt service is senior to dividend payment, which may result in poor or no dividends to shareholders.

Companies in Britain are not legally classified as monopolies until they own 25% of the market in which they trade. If one assesses all the major sectors in which superstores trade, then Tesco, the largest, emerges with 17% (twice as high as two years ago), and Sainsbury has 13%. If, on the other hand, you assess the sales of groceries, then Tesco emerges with 25% and Sainsbury with 20%.

Hopes that Internet shopping would provide opportunities for new companies to challenge the dominance of the big stores have also been banished. Tesco, the market leader in the grocery business, has already emerged as the biggest online grocer in the world. At the beginning of 2000, it boasted annual Internet sales of £125 million and claimed it would treble that number by the end of the year. In this example, Tesco took the risk of developing a new market long before its competitors identified the benefits of Internet shopping.

Some analysts have argued that Britain's biggest chains collectively meet the legal definition of a monopoly. The five biggest supermarket chains sell 74.5% of all groceries sold in Britain. This could be the most concentrated market on earth, and is seen by many as a cartel which sets the prices of groceries and thus reduces the risks of competition from smaller organisations in the grocery market. Their profits have long been higher than those of similar chains anywhere in continental Europe (Monbiot, 2000).

The four large UK banks, Barclays, HSBC, Lloyds TSB and Royal Bank of Scotland, control approximately 85% of small-business banking. These banks are currently being investigated by the competition commission and face the risk of being fined for fixing charges to customers, thus reducing competition.

Outsourcing is a major tool by which corporations and SBUs relieve risks. Many businesses transfer risk by outsourcing specific activities to other parties. A major supermarket chain, for example, often outsources the storage, quality checks, security and transport of its grocery items to the supplier as a method of transferring risks that are outside its control.

Corporate manslaughter

The current situation is that companies should be prosecuted and convicted for the same general offences as individuals and be subject to the same general rules for the construction of criminal liability. The law should recognise and give effect to the widely held public perceptions that companies have an existence of their own and can commit crimes as entities distinct from the personnel comprising the company. The best method of assessing whether a company possesses the requisite degree of blameworthiness is through adoption of the corporate *mens rea* doctrine. While this inevitably will raise problems of how to assess policies and procedures to ascertain whether they reflect the requisite culpability, such a task is not impossible.

The message is clear; there is now a momentum, fuelled by strong public opinion in the wake of recent disasters, for companies and their directors to be held accountable when death and serious injury occur owing to their perceived failures. In the wake of these events, corporations are subject to new risks and must therefore incorporate sufficient guidelines into their health and safety legislation.

Consolidation

In seeking to reduce risk, opportunities for privatisation are now more limited than in the mid-1980s, however, as the most accessible possessions of the state have already been procured, and there is public resistance to more ambitious schemes. Now many of the larger corporations have chosen a new route to growth: consolidation. By engineering a single, harmonised global market, in which they can sell the same product under the same conditions anywhere in the world, they are looking to extract formidable economies of scale. They are seizing, in other words, those parts of the world that are still controlled by small and medium-sized businesses. Decisions associated with investments on a global basis must take into consideration the country risk analysis.

Consolidation in the print and broadcast media industries has also enabled a few well-placed conglomerates to exert a prodigious influence over public opinion. They have used it to campaign for increased freedom for business. Globalisation, moreover, has enabled companies to hold a gun to government's head. Governments refusing to meet corporate demands will be threatened with disinvestment, or shifting the whole operation to different countries, such as Thailand, resulting in wide-scale unemployment. The result is unprecedented widespread power for corporate bodies (Monbiot, 2000).

Oil companies often suffer from cash flow risk when crude oil prices fall because their cash flows are based on higher crude oil prices. The risk associated with crude oil prices is normally outside the control of the oil companies and can often result in projects being delayed or in decreased output.

Corporate risk strategy

Corporate risk strategy often entails planned actions to respond to identified risks. A typical corporate risk strategy includes:

- accountabilities for managing the corporate risk
- a corporate risk register, which will be maintained as a record of the known risks to the corporate strategy plan, the types of mitigating actions that can then be taken, and the likely results of the mitigating action

- identification of treatment plans that form part of the corporate strategy and will be communicated to the SBUs, so they in turn may manage that risk which may affect them.

A first estimate of potential effects can be determined using assumption analysis, decision tree analysis and the range method. These models can then be used to evaluate the effectiveness of potential mitigating actions and hence select the optimum response. Chapman and Ward (1997) believe mitigating actions can be grouped into four categories, and potential action includes:

- risk avoidance:
  - cancel a project
  - move out of a market
  - sell off part of the corporation
- risk reduction:
  - acquisitions or mergers
  - move to the new market
  - develop a new product/technology in an existing market
  - business process re-engineering
  - corporate risk management policy
- risk transfer:
  - partnership
  - corporate policy on insurance
- risk retention:
  - a positive decision to accept the risk owing to the potential gain it allows.

Many of the mitigating actions at the corporate level generate (or cancel) individual projects or entire programmes conducted at lower levels.

Corporate risk: an overview

Most failures are caused almost exclusively by human failure and by an absence of satisfactory risk management controls. For example, the recent terrorist attack on the twin towers in New York was an unforeseen event; however, the risk management team should have taken measures to evacuate personnel in the event of a terrorist attack on the basis of the data held by US Government agencies.

The worrying fact for senior managers of all types of companies is that the potential for corporate disaster on a large scale is growing at an alarming rate, and worse still, the spectre of the corporate Armageddon is growing at a faster rate than the ability of most organisations to cope. History shows that corporate vulnerability is mainly due to human error. Avoidance of these risks can be achieved by comparing old, painful risks with some new, excruciating ones. Only 15 years ago, the majority of risks faced by firms in the UK were related to day-to-day operations. The most obvious were physical, including standard property risks such as fire and theft of plant and machinery, and human, including standard liability risks such as injury to the workforce or customers. These risks still exist today and have not diminished in significance, but many forward-thinking firms are now willing and able to retain a much higher level of mainly 'attritional' risks, which helps them focus attention on a whole host of new risks of an altogether more complex and unpleasant nature.

Strategic business management

Langford and Male (2001) define an SBU as follows:

> Large firms will normally set up a strategic business unit. It will have the authority to make its own strategic decisions within corporate guidelines, that will cover a particular product, market, client or geographic area.

Strategy is a set of rules which guide decision makers about organisational behaviour, which go on to produce a common sense of direction.

Strategic business management functions

In general, the roles and responsibilities for strategic business managers are as follows:

- They are responsible for managing and coordinating various issues at strategic business level, and for ensuring coherency and conformity to the corporate strategy implementation plan as well as the strategic business plan.
- They will be concerned with macro aspects of the business; these include:
  - political and environmental issues
  - finding a niche in the market and exploiting it
  - business development
  - sustainability or long-term goals of the strategy
  - stakeholders' satisfaction
  - long-term demands of customers or end-users
  - identifying and responding to strategic business risks.

In terms of legal foci, the strategic business manager will abide by planning regulations, environmental restrictions and British Standards. At strategic business level the manager will look at a wider perspective, for example looking at stakeholder arrangements (balancing equity, bonds, debt, and contractual legal agreements between partners). Business managers ensure that everything conforms to current legislation throughout the strategy. The use of environmental impact assessment at strategy level provides a platform for the public to participate in mitigation decisions. This in turn fosters integrity and coordination and shows the stakeholders the benefits of the strategic business manager.

In terms of risk management, the strategic business manager will need to address all possible risks, mitigate and review, documenting as work progresses. The business manager will be concerned with a wider view of business risks, such as the interdependencies of the projects within the strategy, the overall financial risks of the projects, and risks posed by delays in completion of tasks and by sudden changes due to external influences.

In terms of schedule and cost, the strategic business manager will have to look at the whole picture, where comparisons can be made between different projects. The business manager will be concerned with predicting overall profit and loss forecasts within the business level and with long-term profitability, as well as realisation of the business strategy benefits. Strategic business managers coordinate the interface of the projects within the strategy, the coordination logistics, both in design and in the implementation stages. They also consolidate and analyse changes with respect to the overall impact on the business strategy plan and cost.

Business strategy

Corporate strategy is concerned with the company as a whole, and for large, diversified firms it is concerned with balancing a portfolio of businesses, different diversification strategies, the overall structure of the company and the number of markets or market segments within which the company competes (Langford and Male, 2001).

Business strategy, however, is concerned with competitiveness in particular markets, industries or products. Large firms will normally set up an SBU with the authority to make its own strategic decisions within corporate guidelines that will cover a particular product, market, client or geographic area. Finally, the operating or functional strategy is at a more detailed level and focuses on productivity within particular operating functions of the company and their contribution to the corporate whole within an SBU.

An organisation's competitive business strategy is the distinctive approach taken at business level when positioning itself to make the best use of its capabilities and stand out from competitors. From the work of Porter on corporate strategy (1970–2002), the author has developed four key elements that determine the limits of competitive strategy at business level. These are divided into internal and external factors. Internal factors include the organisation's strengths and weaknesses, and the values of key implementers at strategic business level. External factors include business opportunities, threats and technology advances, and expectations of the business environment within which the organisation operates.

Porter believes an organisation's strategy is normally defined by four components:

- *Business scope.* The customers/end-users served, their needs and how these are being met.
- *Resource utilisation.* Resourcing properly the areas in which the organisation has well-developed technical skills or knowledge bases – its distinctive capabilities.
- *Business synergy.* Attempting to maximise areas of interaction within the business such that the effect of the whole is greater than the sum of the parts.
- *Competitive advantage.* Determining these sources.

At the corporate level of the organisation, senior managers will develop a corporate strategy that is concerned with balancing a portfolio of businesses. Corporate strategy is company-wide and is concerned with creating competitive advantage within each of the SBUs. Business strategy is concerned with which markets the firm should be in and transferring the relevant information to corporate level. The divisionalised structure, as part of the whole portfolio of businesses, will have different strategic time horizons for each division that has to be incorporated by the main board to produce an integrated corporate strategy.

Many SBUs need to borrow money to finance projects. Lenders often require parent company guarantees from the corporation in case of default by the SBU. SBUs will, in some cases, use the corporation's profit and loss accounts as a means of illustrating their financial stability to clients rather than their own accounts, which are often not as financially sound.

Figure 9.2 illustrates the relationship between the SBUs and the corporate and project levels. SBUs are seen to be subordinate to the corporate entity but senior to projects in diverse business sectors whilst remaining under the corporate umbrella.

Each business unit must submit a summary of its proposed strategies and business plans to the corporate board. This is called the five-year commitment (FYC). The combined FYCs of all the businesses must achieve the corporate objectives. The FYC is a five-year business plan which is updated each year and moved forward by a

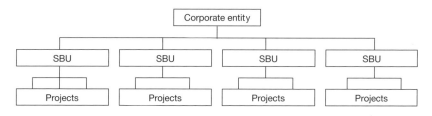

Figure 9.2. Typical SBU organisation

year. The SBUs will update or add more issues and commitments and will include a business risk register covering similar points to that of the corporate risk strategy.

The tools and information used at SBU level are similar to those used at corporate level. The business unit strategy, derived from the corporate strategy, is still concerned with survival and increasing value but is focused on its particular market area, normally a portfolio of projects.

Focusing on the difference, the owner comes from the SBU and the champion is a senior executive with regular contact with the corporate board. It is now more important that the core senior executives and project managers consider input from the customers, partners and suppliers as that interface is much closer. Major decisions must be ratified through regular contact with the corporate board.

The scope is focused on the market of the SBU but extends beyond the current project portfolio, looking for new opportunities. It now includes review and control of individual projects, as well as compliance with corporate strategy decisions.

Much of the same information is used, but it focuses in greater detail on the particular market area. The same identification tools are appropriate, PEST and SWOT. In addition, health and safety management and environmental management systems will identify some risks that are generic to all projects in that market area, particularly those associated with production processes and methods, such as chromium plating, removal of toxic waste and working conditions.

Today's marketplace demands cost-effectiveness, competitiveness and flexibility from a business if it is to survive and grow. Such demands necessitate effective business plans, both strategic in support of longer-term goals and tactical in support of ever-changing business needs and priorities and their associated risks.

A critical factor in this is the synergy required between business operations and associated information systems and technology architectures. A further key factor is understanding and dealing with the legislative, environmental, technological and other changes that impact on an organisation's business.

Project management

A project is a unique investment of resources to achieve specific objectives. Projects are realised to produce goods or services in order to make a profit or to provide a service for the community. The project itself is an irreversible change, with a project life cycle and a defined start and completion date. Any organisation has an ongoing line management of the project requiring management skills. According to the Association for Project Management (Simon *et al.*, 1997),

> Project Management is the planning, organisation, monitoring and control of all aspects of a project and the motivation of all involved to achieve project objectives safely and within defined time, cost and performance.

Turner (1994) presents quite a rosy future for project management, and recognises the changes it is currently undergoing and that it is likely to continue to see in the years ahead. In this challenging and ever-changing environment, project management has emerged as a discipline that can provide the competitive edge necessary to succeed, given the right manager.

The new breed of project manager sees a natural salesperson who can establish harmonious customer relations and develop trusting relations with stakeholders. In addition to some of the obvious keys to project managers' success – personal commitment, energy and enthusiasm – it appears that, most of all, successful managers must manifest an obvious desire to see others succeed.

The project manager's responsibilities are broad and fall into three categories: responsibility to the parent organisation, responsibility to the project and the client, and responsibility to members of the project team. Responsibility to the SBU itself includes proper conservation of resources, timely and accurate project communication and careful, competent management of the project. It is very important to keep senior management of the parent organisation fully informed about the project's status, cost, timing and prospects. The project manager should note the chances of being over budget or being late, as well as methods available to reduce the likelihood of these dread events. Reports must be accurate and timely if the project manager is to maintain credibility, protect both the corporate body and SBUs from high risk, and allow senior management to intercede where needed.

Communication is a key element for any project manager. Running a project requires constant selling, reselling and explaining the project to corporate and SBU levels, top management, functional departments, clients and all other parties with an interest in the project, as well as to members of the project team itself. The project manager is the project liaison with the outside world, but the manager must also be available for problem solving, and for reducing interpersonal conflict between project team members. In effect the project manager is responsible to all stakeholders regarding the project to be managed.

Control of projects is always exercised through people. Senior managers in the organisation are governed by the CEO, who is directed by such groups as the executive committee and/or the board of directors. Senior managers, in turn, try to exercise control over project managers, and the project managers try to exert control over the project team. Because this is the case, there is a certain amount of ambiguity, and from time to time humans make mistakes. It is therefore important that there are effective communication controls, standards and procedures to follow.

According to Turner and Simister (2000), the roles and responsibilities for project managers are as follows:

- They are responsible for managing and coordinating various issues at project level, and for ensuring coherency and conformity to the project strategy implementation plan by working hand in hand with the strategic business manager.
- They will be more project focused. For example, they will be concerned with micro aspects of each project in question, such as the mechanics of delivery of a single project to timescales, cost budgets and quality of deliverables.
- In terms of legal foci, the project manager will abide by planning regulations, environmental restrictions and British Standards.
- Here the project manager will adopt the standard legal requirements specified at the business level but tailor these requirements to suit each project.
- In terms of risk management, the project manager will need to address all possible risks, mitigate and review, documenting as work progresses.
- The project manager will assess risks in the individual projects, but will report to the business manager on the next level if significant impact on the overall strategy and cost is foreseen.
- In terms of schedule and cost, the project manager will have to look at the individual project, and use the tools and techniques available to analyse the project.

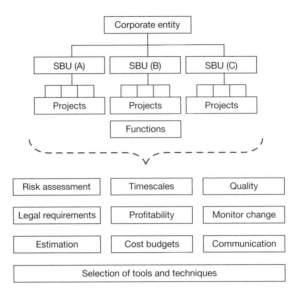

Figure 9.3. Typical project management functions

- The project manager will be concerned with individual profitability.
- They will coordinate the interfaces of the individual project stages.
- The work should be completed within cost, time and quality restraints.
- Cost plan and cost control to meet the allocated budget for each project.
- They should monitor changes and report them to business level if necessary.

Figure 9.3 illustrates the typical project management functions carried out at project level by the project management team. The different functions are often dependent on the type of project undertaken, and these are often monitored by the SBU as well as the project manager.

In the world of project management, it has been common to deal with estimates of task durations and costs as if the information were known with certainty. On occasion, project task workers have inflated times and costs and deflated specifications on the grounds that the project manager or SBU manager would arbitrarily cut the project budget and duration and add to the specifications, thereby treating the problem as a decision under conflict with the management as an opponent.

In fact, a great majority of all decisions made in the course of managing a project are actually made under conditions of uncertainty. In general, many project managers adopt the view that it is usually best to act as if decisions were made under conditions of risk. This will often result in estimates being made about the probability of various outcomes. If project managers use appropriate methods to do this, they can apply knowledge and skills they have to solving project decision problems.

Project risk management is a process which enables the analysis and management of risks associated with a project. Properly undertaken, it will increase the likelihood of successful completion of a project to cost, time and performance objectives. However, it must be noted that no two projects are the same, causing difficulties with analysis and troubleshooting. In most cases things go wrong that are unique to a particular project, industry or working environment. Dealing with project risks is therefore different from situations where there is sufficient data to adopt an actuarial approach.

The first step at project level is to recognise that risk exists as a consequence of uncertainty. In all projects, there will be risks of various types:

- a technology is yet to be proven
- lack of resources at the required level
- industrial relations problems
- ambiguity within financial management.

Project risk management is a process designed to remove or reduce the risks which threaten the achievement of project objectives. It is important that management regard it as an integral part of the whole process, and not just simply a set of tools and techniques.

Why project risk management is used

There are many reasons for using project risk management, but the main reason is that it can provide significant benefits far in excess of the cost of performing it.

Turner and Simister (2000) believe benefits gained from using project risk management techniques serve not only the project but also other parties, such as the organisation as a whole and its customers. Below is a list of the main benefits:

- There is an increased understanding of the project, which in turn leads to the formulation of more realistic plans, in terms of cost estimates and timescales.
- Project risk management gives an increased understanding of the risks in a project and their possible impact, which can lead

to the minimisation of risks for a party and/or the allocation of risks to the party best suited to handle them.

- There will be a better understanding of how risks in a project can lead to a more suitable type of contract.
- It will give an independent view of the project risks, which can help to justify decisions and enable more efficient and effective management of risks.
- It gives knowledge of the risks in projects, which allows assessment of contingencies that actually reflect the risks and which also tends to discourage the acceptance of financially unsound projects.
- It assists in the distinction between good luck and good management, and bad luck and bad management.

Beneficiaries from project risk management include:

- Corporate and SBU senior management, for whom a knowledge of the risks attached to proposed projects is important when considering the sanction of capital expenditure and capital budgets.
- The client, as the client is more likely to get what it wants, when it wants it and for a cost it can afford.
- The project management team, who want to improve the quality of their work. It will help meet project management objectives such as cost, time and performance.
- Stakeholders in the project or investment.

Project risk management should be a continuous process that can be started at any early stage of the life cycle of a project and can be continued until the cost of using it is greater than the potential benefits to be gained. It will be far more effective to begin project risk management at the start of a project because the effects of using it diminish as the project travels through its life cycle.

Many project management procedures place considerable stress on the quantification of risk, although much evidence suggests that this is erroneous as many top executives ignore data in favour of intuition (Traynor, 1990). The emphasis placed on the quantification processes fails to prompt a manager to take account of other areas more difficult or impossible to quantify, thus excluding a large element of risk.

It would be of great help if one could predict with certainty, at the start of a new project, how the performance, time and cost goals would be met. In some projects it is possible to generate reasonably accurate predictions; however, the larger the project is, often the less accurate these predictions will be. There is considerable uncertainty about organisations' ability to meet project goals.

Uncertainty decreases as the project moves towards completion. From the project start time, the band of uncertainty grows until it is quite wide by the estimated end of the project. As the project develops, the degree of uncertainty about the final outcome is reduced. In any event, the more progress made on the project, the less uncertainty there is about achieving the final goal.

The project manager must have good knowledge of the stakeholders in the project and their power. A consensus must be found with the majority of participants in the project. This is often not easy, because stakeholders have conflicting interests. It is important that project managers continuously analyse the positions of the stakeholders, their expectations, their needs and their foreseeable reactions. If the stakeholders think that they will only be collaborating once, then it is difficult to achieve creative cooperation.

Risks at project level

A project manager must cope with different cultures and different environments. Different industries have different cultures and environments, as do different regions and countries. The term 'culture' refers to the entire way of life for a group of people. It encompasses every aspect of living and has four elements that are common to all cultures: technology, institutions, language and arts (Turner and Simister, 2000).

The technology of a culture includes such things as tools used by people, the material things they produce and use, the way they prepare food, their skills and their attitude towards work. It embraces all aspects of their material life (Haynes, 1990).

The institutions of a culture make up the structure of society (*The Economist*, 2001). This category contains the organisation of the government, the nature of the family, the way in which religion is organised as well as the content of religious doctrine, the division of labour, the kind of economic system adopted, the system of education, and the way in which voluntary associations are formed and maintained.

Language is another ingredient of all cultures. The language of a culture is always unique because it is developed in ways that meet the express needs of the culture of which it is part. The translation of one language into another is rarely precise. Words carry connotative meanings as well as denotative meanings. The word 'apple' may denote a fruit, bribery ('for the teacher'), New York City, a colour, a computer, favouritism ('of my eye'), and several other things (Johnson and Scholes, 1999).

Finally, the arts or aesthetic values of a culture are as important to communication as the culture's language. If communication is the glue that binds culture together, art is the most important way of communicating. Aesthetic values dictate what is found beautiful and satisfying. If a society can be said to have style, it is from culture's aesthetic values that style has its source (Jaafari, 2001).

The project audit is a thorough examination of the management of a project, its methodology and procedures, its records, its properties, its budgets and expenditures, and its degree of completion. It may deal with the project as a whole, or only with a part of the project. The formal report should contain the following points:

- *Current status of the project.* Does the work actually meet the planned level of completion?
- *Future status.* Are significant schedule changes likely? If so, the nature of these changes should be indicated.
- *Status of crucial tasks.* What progress has been made on tasks that could decide the success or failure of the project?
- *Risk assessment.* What is the potential for project failure or monetary loss?
- *Information pertinent to other projects.* What lessons learned from the project being audited can be applied to other projects being undertaken by the organisation?
- *Limitations of the audit.* What assumptions or limitations affect the data in the audit?
- Tools and techniques used at project level.

One must note that the project audit is not a financial audit. The project audit is much broader in scope and may deal with the project as a whole or any competent set of components of the project. It may be concerned with any part of project management.

While the project audit will be concerned with all aspects of project management, it is not a traditional management audit. Management audits are primarily concerned that the organisation's management systems are in place and operative. The project audit goes beyond this. Among other things, it is meant to ensure that the project is being appropriately managed. Some managerial systems apply fairly well to all projects, for example the techniques of planning, scheduling, budgeting and, of course, risk management (Turner and Simister, 2000).

The author also believes that decommissioning risks play a fundamental part in risk management at project level. These are the risks associated with plant or machinery at the end of the project life cycle. For example, what will be done to a nuclear power station when it is decommissioned? What are the costs of decommission? What are the environmental effects? And which stakeholders are affected and how?

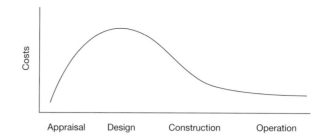

Figure 9.4. The project risk cycle (Merna and Owen, 1998)

Cooper and Chapman (1987) suggest that the need for emphasis on risk assessment is particularly apparent when projects involve:

- large capital outlays
- unbalanced cash flows requiring a large proportion of the total investment before any returns are obtained
- significant new technology
- unusual legal, insurance or contractual arrangements
- important political, economic or financial parameters
- sensitive environmental or safety issues
- stringent regulatory or licensing requirements.

A combination of a number of, the above parameters are fundamental to project strategies. Each risk identified in the project must have a uniform basis of assessment, which will inevitably involve cost and time.

A typical project risk cycle for a construction project is shown in Figure 9.4, where the level of risk is plotted against the stage of the project. As the diagram indicates greater risk at the earlier stages of the project cycle, it can be concluded that this is where the majority of risk management efforts should be concentrated, as it offers greater yields (Merna and Owen, 1998).

Risk management is used throughout the full life cycle of the project from pre-tender through to after-market.

The risk management plan is the process of identifying and controlling business, technical, financial and commercial risks throughout the project's life cycle by eliminating or reducing the likelihood of occurrence and potential impact of the threat. For commercial undertakings, any impact on the project outcomes needs to be expressed in terms of cost. Financial impact is therefore a baseline from which to measure risk. Risk that has a timescale is to be converted into cost. This will enable accounts to raise provisions early in the project if they are needed.

The risks associated with individual projects carried out at project level must be addressed at the SBU level. SBUs deal with a portfolio of individual projects, each having its own risk characteristics. SBUs need to address the financial implications associated with a portfolio of projects and not just on an individual-project basis.

Corporate, strategic business, and project risk assessment

Within any organisation performing risk management, tools and techniques must be used at each level (as discussed in Chapter 6). The use of these tools and techniques allows data to be analysed and this analysis provides the basis of whether or not an investment takes place. Stakeholders are also identified at each level, and are allowed to contribute to the risk management process. These stakeholders must be identified and their requirements recorded, as well as their relative significance. In order to assess the risks at each level, various tools and techniques may be applied. These techniques may generally be applied at each level in the process, but some will be more pertinent to a particular level than others. Figure 9.5 illustrates the levels and the required input at each level. The tools and techniques used at each level will be determined by the risk assessor, since no specific techniques(s) can be identified at each level.

Figure 9.5 divides the organisation into corporate, strategic business, and project levels. At each level, risk management tools and techniques are used and stakeholder inputs are taken into consideration. This process forms a basis for the risk management system.

Figure 9.6 illustrates the risk management process, which includes the identification, analysis and control of risks to be applied at corporate, strategic business, and project levels. The risk

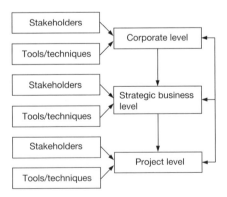

Figure 9.5. Risk management mechanism

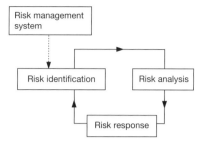

Figure 9.6. Risk management process

management system is dynamic and must be continuous over the project life cycle.

Information which forms the basis of this risk management mechanism is gathered from a literature review and a questionnaire sent out to the FTSE-listed non-construction and construction organisations.

At each level within the organisation, the author proposes a generic system, illustrated in Figure 9.7, with the purpose of identifying, analysing and responding to risks specific to each level within the organisation.

This process illustrated in Figure 9.7 should be a dynamic process carried out throughout the whole project life cycle in a continuous loop.

Figure 9.7 illustrates the processes the author believes should be undertaken at each level of an organisation, the stakeholders and risk management tools and techniques being involved as and when deemed necessary.

Firstly, the project should be defined. It is imperative that the business or project objectives are identified and clearly understood by the project team prior to embarking upon risk management activities. At this stage each level of the organisation should define what the project means to them, for example business or project requirements, client specification, work breakdown structure, cost estimates, project programme, cost of finance, constraints and project implementation plan. This is done through the use of historical data, organisational specific knowledge, and information specific to the project in hand and the corporation's overall goals, normally associated with the five-year plan.

Once the project has been defined, at each level, the risk identification process begins. The process of identifying risks is carried out through the use of a variety of techniques suited to the type of

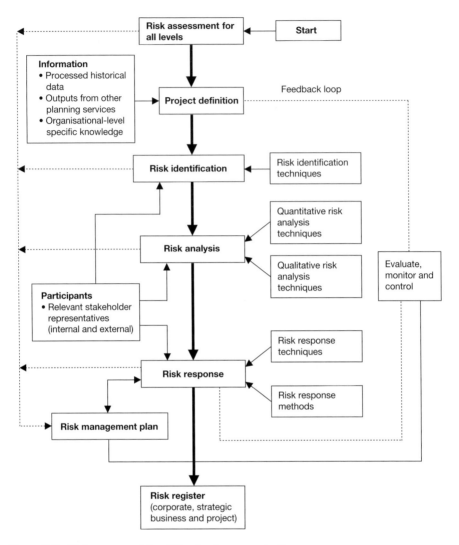

Figure 9.7. Risk assessment for all levels of an organisation

project and the resources available (as discussed in Chapter 4). The allocation of risk owners/facilitators is undertaken during this stage, which aims to allocate ownership of risk to the person best placed to control and manage it. Identified risks and risk owners will be recorded on the risk register, which later will become a database at SBU level.

The information gathered at the identification stage is then analysed. Risk analysis tools and techniques, either qualitative or quantitative, are now employed to provide a thorough analysis of the risks specific to the project at each level within the organisation. Analysis may include defining the probabilities and impacts of risk and the sensitivity of the identified risks at each level.

After completion of the identification and analysis processes, response to these risks can be carried out. This process is carried out through the use of risk response methods and techniques. If the response decision is to mitigate the risks, the costs of mitigation must be assessed extensively and budgeted for accordingly. Retained risks at each level will be put on the risk register and be constantly reviewed.

Within this model, stakeholders are of particular importance. Stakeholders are involved at each level and will have input at each stage in the risk assessment process (identification, analysis and response). The model allows information from each stage to flow backwards and forwards through the organisation, where it can then be continually monitored, evaluated and controlled.

Once all the information has been processed through the model, a risk management plan is constructed and implemented. The plan should form an integral part of project execution and should give consideration to resources, roles and responsibilities, tools and techniques, and deliverables. This plan will include a review of the risk register, monitoring of progress against risk actions and reporting. The final output of the model will be a risk register at the corporate, strategic business, and project levels.

Feedback is a key vehicle used in this proposed mechanism for the organisation to learn from both its successes and mistakes, internally or externally. It provides continuous improvement at both SBU and project levels, and risk management itself.

Feedback is a continual process, gathering data from known and unforeseen events. Information is held at SBU level and disseminated throughout the organisation.

These risk assessments and risk registers at the corporate, strategic business, and project levels will be made available to each level within the organisation.

An overall risk register, incorporating risk registers developed at corporate, strategic business, and project levels will be developed at strategic business level and be continually updated as the project develops. It is important that the risk assessments carried out for the projects at SBU level are all in the same format, forming a formal database for all projects. This will allow data to be obtained for any project in the same format so comparisons can be made.

Risk assessment at corporate, strategic business, and project levels should run concurrently. At any time during the assessments, risks

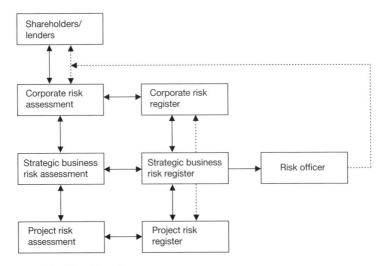

Figure 9.8. The risk cycle

that may result in the project or investment not being sanctioned can be flagged up from any level.

The proposed risk management mechanism used will:

- identify and manage risks against defined objectives
- support decision-making under uncertainty
- adjust strategy to respond to risk
- maximise chances through a proactive approach
- increase chances of project and business success
- enhance communication and team spirit
- focus management attention on the key drivers of change.

Figure 9.8 illustrates the interaction of the different levels within the organisation. Information regarding risk assessment and risk registers is passed freely through the organisation.

Within this proposed mechanism, the strategic business level will act as the conduit between the corporate and project levels. A risk officer will be designated at the strategic business level with responsibilities for ensuring that risks managed at both corporate and project levels are registered and that any further risks identified will be incorporated in the risk register held by the risk officer. All the information gathered from the corporate, strategic business, and project levels will be collated and passed on to the risk officer. The risk officer will be in direct contact with risk facilitators at the corporate and project levels. This mechanism will ensure that all levels of the organisation will have an input into the overall risk register.

Managers and owners of risks retained and mitigated will be in the corporate, strategic business or project level within the organisation, depending where the risk originates. For example, a risk originating at project level will be managed and owned by the project manager. The risk assessments and risk registers held by the project manager will be passed to the risk officer, at strategic business level. The risk officer will review the overall register and inform both the corporate and the strategic levels of any changes in risk assessment as the project proceeds.

The advantages of the strategic business level of an organisation holding a risk register as a conduit from both the corporate and the project levels include the following.

- The strategic business level is immediate to both the corporate and the project levels.
- One risk officer is responsible for all the risks.
- If any information is required about risk specific to a project, both the project and the corporate levels will know where to get this information from.
- Both the project and the corporate levels will have access to all risk management systems and information.
- Stakeholders will have easy access to how risks are managed at all levels of an organisation.
- Risks throughout the organisation are controlled by one experienced and qualified department.
- The risk officer will have access to other projects being carried out at the SBU level.

However, in order for the process to work, regular reviews need to take place. These may be in the form of risk workshops involving risk facilitators from the corporate and project levels in conjunction with the risk officer. Stakeholders at the corporate and project levels such as shareholders and suppliers will coordinate with the designated risk facilitator at the respective levels on a regular basis and report to the risk officer.

New risks, the cost of managing such risks and the status of all the risks identified at each level will be addressed in the overall risk register managed by the risk officer at SBU level.

Summary

This chapter outlines the functions of the corporate, strategic business, and project levels in a typical corporation. Each level is responsible for managing the risks identified and ensuring that identified risks and their management are made available to all other levels. In most cases risks are typical to each level. Corporate risks tend to be associated with political, legal, environmental and finance elements of an investment. Many of these risks are assessed in greater detail at strategic business level. Project risk management often entails risks being assessed in greater detail, as more data are available and risks become more specific to the project rather than to the overall strategic risk assessed at corporate level. To ensure all risks at all levels are managed, it is paramount that an overall risk management system is implemented and that risks identified at all levels are managed over the project life cycle.

Bibliography

Cadbury, A. Report of the Committee on Financial Aspects of Corporate Governance. Gee, London, 1992.

Chapman, C. B. and Ward, S. C. *Project Risk Management: Processes, Techniques and Insights.* Wiley, Chichester, 1997.

Cooper, D. and Chapman, C. *Risk Analysis for Large Projects: Models, Methods and Cases.* Wiley, Chichester, 1987.

David, F. R. How companies define their mission. *Long Range Planning,* **22**(1) (1989), 90–97.

Falsey, T. A. *Corporate Philosophies and Mission Statements.* Quorum, New York, 1989.

Grundy, T. Strategy implementation and project management. *International Journal of Project Management,* **16**(1) (1998), 43–50.

Handy, C. *Beyond Certainty. The Changing Worlds of Organisation.* Harvard Business School Press, Harvard, 1999.

Haynes, M. E. *Project Management: from Idea to Implementation.* Kogan Page, London, 1990.

Hussey, D. E. *Introducing Corporate Planning – Guide to Strategic Management,* chapter 1. Pergamon, London, 1991.

Institute of Chartered Accountants. Internal Control: Guide for Directors on the Combined Code. [Turnbull Report.] ICA, London, 1999.

Jaafari, A. Management of risks, uncertainties and opportunities on projects: time for a fundamental shift. *International Journal of Project Management,* **19** (2001), 89–101.

Johnson, G. and Scholes, K. *Exploring Corporate Strategy,* 4th edition. Prentice-Hall Europe, 1999.

Keasey, K., Thompson, S. and Wright, M. *Corporate Governance: Economic. Management and Financial Issues.* Oxford University Press, Oxford, 1997.

Langford, D. and Male, S. *Strategic Management in Construction.* Blackwell Science, Oxford, 2001.

McCoy, C. S. *Management of Values.* Pitman, Boston, 1985.

Merna, A. and Owen, G. *Understanding the Private Finance Initiative. The New Dynamics of Project Finance.* Asia Law and Practice, Hong Kong, 1998.

Monbiot, G. *Captive State. The Corporate Takeover of Britain.* Pan, London, 2000.

Reichmann, P. Profile business. *Sunday Times,* 7 Mar., 1999, section 3, p. 6.

Simon, P., Hillson, D. and Newland, K. (eds). *Project Risk Analysis and Management Guide.* Association for Project Management, High Wycombe, 1997.

Spencer, A. *On The Edge of the Organisation: the Role of the Outside Director.* Wiley, New York, 1983.

The Economist. Risk new dimension. 29 Sept., 2001.

Traynor, V. T. Project risk analysis. MSc thesis, UMIST, Manchester, 1990.

Tricked, R. I. *International Corporate Governance: Text, Reading and Cases,* p. 9. Prentice-Hall, Singapore, 1994.

Turner, J. R. Project management: future development for the short and medium term. *International Journal of Project Management,* **12**(1) (1994), 3–4.

Turner, R. and Simister, J. *The Handbook of Project Management.* Gower, Aldershot, 2000.

Yoshimori, M. Whose company is it? The concept of the corporation in Japan and in the West. *Long Range Planning,* **28** (1995), 33–45.

Zahra, S. A. and Pearce, J. A. Board of directors & corporate financial performance: a review and integrative model. *Journal of Management,* **15** (1989), 291–334.

CHAPTER TEN

Developments in risk management

N. J. Smith

This chapter consolidates the progress of the process of risk manage-ment and speculates on its continuing evolution. The nature of risk itself indicates that trend analysis will, at best, only be relevant for a proportion of events and that it is likely to be incomplete and incon-sistent in its approach. Nevertheless, the chapter commences with a review of recent achievements in construction risk management. It then progresses to consider the role of risk in a sustainable environ-ment. With increasing populations and demands for higher levels of infrastructure on a planet with finite resources, this is one of the great challenges of tomorrow. The chapter concludes with some brief speculations on the future of risk management.

Trends in risk management

As shown in the early chapters of the book, the nature and under-standing of risk management is not aided by the terminology and jargon employed by professionals and analysts in different industrial sectors. However, as new codes of practice develop and standardised definitions become accepted, the effect of these variations will be reduced considerably.

The nature of risk management is, at one and the same time, fixed and known yet also dynamic and continually evolving. To quote Alphonse Karr in '*Les Guepes*' of 1849, it is a case of 'Plus ça change, plus c'est la même chose'. The basic structure for risk management proposed almost 40 years ago by Perry and Thompson and outlined in Chapter 6 has remained substantially unchanged. Most project risk guidebooks have made some minor modifications to the struc-ture, but the essence of the three key stages of 'identify', 'assess' and 'manage' is retained. This can be shown in the RAMP guide approach outlined in Chapter 7.

The evolving trends are discussed in Chapters 8 and 9. In Chapter 8, the relatively new concept of uncertainty management is intro-duced. The basic principle behind uncertainty management is that

opportunities have to be exploited in addition to risks being mitigated. The evolution is concerned with how these two previously
distinct mindsets could be combined in a single process and what
other parameters, such as project complexity, might be key factors
in the process. At this stage there seem to be two major schools of
thought: one that each project situation is unique and a 'situational'
analysis is required on every individual occasion, and the other that
some model of key parameters and their interactions might be
developed for generic application. As the trend for larger, more
complex infrastructure projects increases and project managers
are involved in the full project life cycle and the management of
funding and change, so the need for a better understanding of
uncertainty management grows.

Chapter 9 deals with a change in responsibility for project risk, for
the duration of the project. At first glance this may not seem a major
change in risk management, but on further investigation the
approaches being adopted in the UK are having a radical effect on
the corporate approach to risk. For the first time the directors of
private companies are shown as being aware of the likely nature of
the project risk and of the types of risk mitigation being considered
prior to commencement. This new interaction between construction professionals and marketing and financial professionals has to
be a step closer to sharing a common understanding of the riskiness
of a project.

This interaction of people and risk methodology is a clear
message from the recent Royal Academy of Engineering reports.
The point is made that 'human error' often accounts for a risk event
occurring. All too frequently this is exacerbated by using professional people in a most inappropriate manner within a particular
process. Ideally all routine tasks would be automated, and only tasks
requiring creativity and/or experience would be taken by people.
Clearly there is still a long way to go before this goal is achieved, but
progress is being made.

The trends show that the more our communications can be
improved, to remove ambiguity and to promote understanding,
then the more likely it is that the decisions taken under conditions
of uncertainty will be better project decisions. Better decisions are
the essence of good risk management.

Sustainability

Sustainability is important for all of us, in the ways in which the
planet is managed. Forecasters and futurologists have made predictions of major difficulties awaiting us caused by shortages of water or

other key minerals. Whilst risk management is not concerned with predicting the future, it can and should take into account the issues related to sustainability.

In the provision of infrastructure, sustainability must be considered at several stages. Initially, there is the site. Is it greenfield or brownfield? Will it cause any problems for future uses of the site? Consideration of the materials – are they renewable? Do they recycle any existing waste material? The construction processes and procedures – are these low-energy solutions? Does the service life of the infrastructure meet sustainability criteria? Can the project be decommissioned safely and effectively? These questions are starting to be considered generally, without forming an integral part of the project risk management process. In future these will have to be considered simultaneously.

A recent example of a sustainable risk analysis for an additional power station in a heavily urbanised and industrialised part of a country indicated that the station could be eliminated altogether if energy-saving light bulbs were used. Further analysis showed that it would be feasible for all supplied premises to be issued with their first set of new bulbs free of charge, offset by eliminating unnecessary power generation from the existing plants. Risk management and, increasingly, uncertainty management can be utilised to make these kinds of contributions to our decision-making.

Risk futures

Risk management is still evolving, yet the demand for its application continues to grow. The increasing demand for infrastructure projects, often by collaborative procurement or public–private partnerships, necessitates the improvement of communications between all the stakeholders: the engineers, analysts, suppliers, vendors, contractors, lawyers, financiers, bankers and others. This is required because it is only through improved understanding of the projects and the risks that better project decisions can be made tomorrow.

With the enormous changes in access to computing hardware that have occurred over the last 40 years and the increases in human computer literacy, it is likely that risk management will be devolved to those involved in project decisions, working on real-time or faster-than-real-time analyses. The role of the specialist analyst is likely to reduce. Better decision-making is likely to include optimal use of human experts, consideration of sustainabilty issues and the better understanding of engineering project management.

It is difficult to say whether a perfect risk management system is possible but for the foreseeable future the need to realise all projects on time, within budget and to the specification will require reliance on improving systems of project risk management.

Bibliography

Turnbull, J. *An Introduction to Three Reports on Risk and Engineering: The Societal Aspects of Risk; Common Methodologies for Risk Assessment & Management; Risk Posed by Humans in the Control Loop.* Royal Academy of Engineering, London, 2000.

Index

Page numbers in *italics* refer to tables and figures.